MODERN MILITARY CADENCE ®
CURRENT OPERATIONS

Distributed by:
BYRRD Enterprises, Inc.
1302 Lafayette Ave.
Alexandria, VA 22308

Produced by:
DUNNIGAN Industries, Inc.
8136 Highlands Dr.
Midland, GA 31820

ISBN 0-9679910-0-5
ISBN 1-886715-44-0
ISBN 978-1-886715-44-8

Manufactured in the Uni
10 9 8 7 6 5 4 3 2 1

INTRODUCTION

Much like the military itself, cadences have changed in recent years. Although there are many, "oldies but goodies," still circulating, there is an abundance of new material that should be shared. Modern Military Cadence® combines the running and marching songs of today with some of the unforgettable cadences of the past. Combined, they offer an uninterrupted supply of military heritage in rhythm.

Because this book was compiled from many different sources, the cadences are in no particular service order; both sections, marching and running are from the various services.

Read them to enjoy and remember. Don't get caught short on your big moment when you are asked to move an element. Be prepared, be professional, and be proud. Know what you are going to say, always start when the left foot hits the ground, and you are on your way to singing Modern Military Cadence®.

MARCHING CADENCE 1

RUNNING CADENCE 150

MODERN MILITARY CADENCE®
CURRENT OPERATIONS

MARCHING CADENCES

THE ARMY SONG

First to fight for the right, and to build the Nation's might.

(TROOPS) And the Army goes rolling along.

Proud of all we have done;
Fight until the battle's won.

(TROOPS) And the Army goes rolling along.

And it's Hi, Hi, Hey, the Army's on its way;
Sound off the cadence loud and strong.

(TROOPS) Two, three, four, where e'er we go,
You will always know
That the Army goes rolling along.

HAIL, O INFANTRY

Hail O, hail O, Infantry
Queen of Battle follow me.
Airborne Rangers the life for me,
That's all I ever want to be .

Got a letter in the mail,
Said go to war or go to jail.
Got a letter in the mail,
Said go to war or go to jail.

O hail, O hail, O Infantry,
Queen of Battle follow me,
Infantry Ranger is the life for me,
That's all I ever want to be.

I used to drive a Chevy truck,
Now I pack it in my ruck.
I used to drive a Chevy truck,
Now I pack it in my ruck.

HURRAH, HURRAH

They issued me an M-16,
HURRAH! HURRAH!
They issued me an M-16,
HURRAH! HURRAH!

They issued me an M-16,
But they forgot the magazine
And we'll all be dead.
By the summer of '96.

They issued me a bayonet,
HURRAH! HURRAH!
They issued me a bayonet,
HURRAH! HURRAH!
They issued me a bayonet,
And I haven't got to use it yet
And we'll all be dead.
By the summer of '96.

They issued me a Claymore Mine,
HURRAH! HURRAH!
They issued me a Claymore Mine,
HURRAH, HURRAH!

They issued me a Claymore Miine,
Now all I do is double time
And we'll all be dead
By the summer of '96.

MARINE BASIC

I took my gal down Mexico way,
Had rum and tequila 'till the room did sway.
She taught me to dance, I taught her to sing,
On the way past the border I gave her a ring.

While her eyes were shining full of lust,
Told her my duty was enlist or bust.
She cried at first, but then was proud,
"Semper Fi," she shouted, clear and loud.

They shaved off my hair, gave me a gun,
Told me the future was nothing but fun.
I loaded the car with all of our stuff,
Never believed good-byes would be that tuff.

Now I'm marching with the proud, the few,
Two more months and basic is through.
My gal is waiting, lonely but proud,
Growing and growing with our first child.

LET THE FOUR WINDS BLOW

Let the four winds blow,
Let 'em blow, let 'em blow.
From the East to the West,
U.S. Army is the best.

Ain't no use in looking down,
Ain't no discharge on the ground.
Ain't no use in looking down,
Ain't no discharge on the ground.

Oh yea, oh yea,
Oh yea, oh yea.
From the East to the West,
Echo Company you're the best.

I used to date a beauty queen,
Now I carry an M-16
I used to date a beauty queen,
Now I carry an M-16.

EVERYWHERE WE GO (NAVY)

Everywhere we go
People want to know,
Who we are,
So we tell them.

We're not the Army,
The back-pack Army.

We're not the Air Force
The low-flying Air Force.

We're not the Marines,
They don't even look mean.

We're not the Coast Guard,
They never work hard.

We are the Navy,
The world's finest Navy.

HEY LOTTY DOTTY

Hey Lotty Dotty,
Let's have a party.
Hey Lotty Dotty,
An Echo Company party.

Sat me down in a barber chair,
Turned around I had no hair,
Sat me down in a barber chair,
Turned around and I had no hair.

Hey Lotty Dotty,
Let's have a party.
Hey Lotty Dotty,
Let's have a party.

The air was cold the ground was wet,
When they put me on that big old jet.
The air was cold the ground was wet,
When they put me on that big old jet.

JUST THE OTHER DAY

Just the other day I heard a Drill Sergeant say,
You wanna be soldiers you gotta do it my way.

D-r-i-i-i-i-i-i-i-i-i-v-e on soldiers,
D-r-i-i-i-i-i-i-i-i-i-v-e on soldiers,
Soldiers driving on, drive on soldiers.

Just the other day I heard a Black Hat say,
You want to be Airborne you gotta do it my way.

D-r-i-i-i-i-i-i-i-i-i-v-e on soldiers,
D-r-i-i-i-i-i-i-i-i-i-v-e on soldiers,
Soldiers driving on, drive on soldiers.

Just the other day I heard the CO say,
You want to be Echo, you gotta do it my way.

D-r-i-i-i-i-i-i-i-i-i-v-e on soldiers,
D-r-i-i-i-i-i-i-i-i-i-v-e on soldiers,
Soldiers driving on, drive on soldiers.

WE ARE, WE ARE

We are, we are, we are Echo Company,
And we like to party.
P-A-R-T-Y-
Party Hearty, Party Hearty all night long.

(TROOPS)
Your left, your left, your left, right,
Get on down.
Your left, your left, your left, right,
Get on down.
Now drop and beat your face,
Echo Company's gonna rock this place.
Ahooa, check it out, check it out,
Ahooa, check it out, check it out.

We are, we are, we are Echo Company,
And we like to do PT.
Push ups, sit ups,
Push ups, sit ups, two mile run.

(TROOPS)
Your left, your left, your left, right,
Get on down.
Your left, your left, your left, right,

Get on down.
Now drop and beat your face,
Echo Company's gonna rock this place.
Ahooa, check it out, check it out,
Ahooa, check it out, check it out.

We are, we are, we are Echo Company,
And we like to body slam.
B-O-D-Y- slam,
Body slamming all night long.

(TROOPS)
Your left, your left, your left, right,
Get on down.
Your left, your left, your left, right,
Get on down.
Now drop and beat your face,
Echo Company's gonna rock this place.
Ahooa, check it out, check it out,
Ahooa, check it out, check it out.

FLYING FREE

Flying free up in the clouds,
An Air Force pilot every proud.

Salute our planes as we zoon by,
We show the world how we can fly.

Moving troops, sending aid,
Firing guns, moving raids.

IN THE NAVY

Air Force says that they fly high,
We say they don't even try.
If you want the very best,
Navy puts you to the test.

Army ain't got nothing new,
All Marines say they're proud and true.
If you want the very best,
Navy puts you to the test.

HEY, you troops get out of bed,
Get in gear into the head.
Get those pea coats off the rack,
Before the dawn can even crack.

I used to stay up drinking wine,
Now all I do is double time.
Bunks and drills ain't no big thing,
As long as I can march and sing.

WAY DOWN IN THE VALLEY

Way down in the valley,
I heard a mighty roar.
It was the roar of Mighty Echo,
Using Alpha for a toy.

Pull out on the throttle,
Give it a little gas.
Move over Awful Alpha,
And let the Mighty Echo pass.

Way down in the valley,
I heard a mighty roar.
It was the roar of Mighty Echo,
Using Bravo for a toy.

Pull out on the throttle,
Give it a little gas.
Move over Baby Bravo,
And let the Mighty Echo pass.

Way down in the valley,
I heard a mighty roar.
It was the roar of Mighty Echo,
Using Charile for a toy.

Pull out on the throttle,
Give it a little gas.
Move over Chicken Charlie,
And let the Mighty Echo pass.

Way down in the valley,
I heard a mighty roar.
It was the roar of Mighty Echo,
Using Delta for a toy.

Pull out on the throttle,
Give it a little gas.
Move over Dummy Delta,
And let the mighty Echo pass.

MARINE TOUGH

Groovy, groovy, groovy,
Tough, tough, tough.
You might beat everybody,
But you can't beat us.

Running, running, running,
Running every day.
We'll come out Number One,
Because we lead the way.

Drilling, drilling, drilling,
Drilling every day.
We'll win the final drill,
Because we lead the way.

I HEAR YOU CALLING

I hear you calling,
Calling to me.
The King of Battle,
Field Artillery.

Your left, your right, now get on up,
Your left, your right, now get on down.
Get on up!
Get on down.
Get on up now!
Get on down.
Your funky, funky, boogaloo,
I can do it, so can you.

I hear you calling,
Calling to me.
The Queen of Battle,
Airborne Infantry.

Your left, your right, now get on up,
Your left, your right, now get on down.
Get on up!
Get on down.
Get on up now!

Get on down.
Your funky, funky, boogaloo,
I can do it, so can you.

I hear you calling,
Calling to me.
Law Enforcement,
Airborne MP.

Your left, your right, now get on up,
Your left, your right, now get on down.
Get on up!
Get on down.
Get on up now!
Get on down.
Your funky, funky, boogaloo,
I can do it, so can you.

I hear you calling,
Calling to me.
Combat Law Enforcement,
Ranger MP.

Your left, your right, now get on up,
Your left, your right, now get on down.
Get on up!

Get on down.
Get on up now!
Get on down.
Your funky, funky, boogaloo,
I can do it, so can you.

WE ARE NAVY

In the snow, the sleet, the rain,
We train, we train, we train.

Sound off, we are (unit)
Sound off, Navy.
Bring it on home now,
We are, we are, Navy.

The first to fight, the last to run,
Because Navy's number one.

Sound off, we are (unit)
Sound off, Navy.
Bring it on home now,
We are, we are, Navy.

We'll win the battle, we'll win the war,
We love this country we're fighting for.

Sound off, we are (unit)
Sound off, Navy.
Bring it on home now,
We are, we are, Navy.
First to fight, last to run,

We are Navy, number one.

Sound off, we are (unit)
Sound off, Navy.
Bring it on home now,
We are, we are, Navy.

Hail, all hail the Company,
The best recruits at R.T.C.

Sound off, we are (unit)
Sound off, Navy.
Bring it on home now,
We are, we are, Navy.

FORT BENNING MORNING

We wake up in the morning,
Drill Sergeants rings the bell.
You think it's easy living,
You'll find it's living hell.

Fort B-e-e-e-e-e-e-e-e-n-i-n-g
Fort B-e-e-e-e-e-e-e-e-n-i-n-g

Early morning while your sleeping,
Drill Sergeants come a creeping all around…
Come a creeping all around…
Kill!

You P.T. every morning,
And this you surely know.
It doesn't make a difference,
In rain, sleet, or snow.

Fort B-e-e-e-e-e-e-e-e-n-i-n-g
Fort B-e-e-e-e-e-e-e-e-n-i-n-g

Early morning while you're sleeping,
Drill Sergeants come a creeping all around…

Come a creeping all around...
Kill!

You learn about your weapon,
You march out on the range.
You come back late that evening,
Your body is all in pain.

Fort B-e-e-e-e-e-e-e-e-n-i-n-g
Fort B-e-e-e-e-e-e-e-e-n-i-n-g

Early Morning while you're sleeping,
Drill Sergeants come a creeping all around.
Come a creeping all around...
Kill!

EVERYWHERE WE GO

Everywhere we go,
People want to know,
Who we are,
So we tell them.

We are Echo,
Mighty, Mighty, the Mighty, Mighty Echo
We're not Alpha,
Awful, awful, Alpha,
We are Echo.

Everywhere we go,
People want to know,
Who we are,
So we tell them.

We are Echo,
Mighty, Mighty, the Mighty, Mighty Echo
We're not Bravo,
Baby, baby, Bravo,
We are Echo.

Everywhere we go,
People want to know,

Who we are,
So we tell them.

We are Echo,
Mighty, Mighty, the Mighty, Mighty Echo
We're not Charlie,
Chicken, Chicken. Charlie,
We are Echo.

Everywhere we go,
People want to know,
Who we are,
So we tell them.

We are Echo,
Mighty, Mighty, the Mighty, Mighty Echo
We're not Delta,
Dumb, dumb, Delta,
We are Echo.

U.S. MARINE PRIDE

The Army and the Navy went to the Persian Gulf,
But they couldn't quite handle what they bit off.

Marines went in to pull them out,
We filled old Saddam full of doubt.

There were a lot of bullets and a lot of dead,
But most of them were in Iraqi Army heads.

I'M NOT

I don't know, but I think I might,
Jump from an airplane while in flight.

Got three kids gonna have three more,
Two on the ground and one in the door.

I'm not the preacher or the preacher's son,
But I'll take the money 'til the preacher
comes.

I'm not the butcher or the butcher's son,
But I'll do the killin' 'til the butcher comes.

I'm not the mama or the mama's son,
But I'll do the lovin' 'til the mama's son comes.

CHESTY PULLER

Chesty Puller was a good Marine,
And a good Marine was he.
He called for his knife
And he called for his gun
And he called for his privates three

Booze, booze, booze, said the privates

Merry men are we
There are none so fair
That they can compare
With Marine Corps Infantry

Chesty Puller was a good Marine,
And a good Marine was he.
He called for his knife
And he called for his gun
And he called for his corporals three

We don't give a damn said the corporals,
Booze, booze, booze, said the privates

Merry men are we
There are none so fair

That they can compare
With Marine Corps Infantry

Chesty Puller was a good Marine,
And a good Marine was he.
He called for his knife
And he called for his gun
And he called for his sergeants three

Get this squad in step said the sergeant,
We don't give a damn said the corporals,
Booze, booze, booze, said the privates

Merry men are we
There are none so fair
That they can compare
With Marine Corps Infantry

Chesty Puller was a good Marine,
And a good Marine was he.
He called for his knife
And he called for his gun
And he called for his louies three

We do all the work said the loies,
Get this squad in step said the sergeant,

We don't give a damn said the corporals,
Booze, booze, booze, said the privates

Merry men are we
There are none so fair
That they can compare
With Marine Corps Infantry

Chesty Puller was a good Marine,
And a good Marine was he.
He called for his knife
And he called for his gun
And he called for his captains three

Shine my boots and brass said the captain,
We do all the work said the loies,
Get this squad in step said the sergeant,
We don't give a damn said the corporals,
Booze, booze, booze, said the privates

Merry men are we
There are none so fair
That they can compare
With Marine Corps Infantry

Chesty Puller was a good Marine,

And a good Marine was he.
He called for his knife
And he called for his gun
And he called for his majors three

Who gonna drive my jeep said the major,
Shine my boots and brass said the captain,
We do all the work said the loies,
Get this squad in step said the sergeant,
We don't give a damn said the corporals,
Booze, booze, booze, said the privates

Merry men are we
There are none so fair
That they can compare
With Marine Corps Infantry

Chesty Puller was a good Marine,
And a good Marine was he.
He called for his knife
And he called for his gun
And he called for his colonels three

Who's gonna answer my phone said the colonels,
Who gonna drive my jeep said the major,

Shine my boots and brass said the captain,
We do all the work said the loies,
Get this squad in step said the sergeant,
We don't give a damn said the corporals,
Booze, booze, booze, said the privates

Merry men are we
There are none so fair
That they can compare
With Marine Corps Infantry

Chesty Puller was a good Marine,
And a good Marine was he.
He called for his knife
And he called for his gun
And he called for his generals three

War, war, war cried the generals,
Who's gonna answer my phone said the colonels,
Who gonna drive my jeep said the major,
Shine my boots and brass said the captain,
We do all the work said the loies,
Get this squad in step said the sergeant,
We don't give a damn said the corporals,
Booze, booze, booze, said the privates

Merry men are we
There are none so fair
That they can compare
With Marine Corps Infantry

YELLOW BIRD

A yellow bird,
With a yellow bill.
Was perched upon,
My window sill.
A yellow bird,
With a yellow bill…

(Troops)
was perched upon my window sill.

I lured him in,
With a piece of bread.
And then I smashed,
His yellow head.
I lured him in,
With a piece of bread…

(Troops)
and then I smashed his yellow head.

I called the doctor,
The doctor said…
Indeed my man,
This bird is dead.

I called the doctor,
The doctor said…

(Troops)
Indeed my man this, bird is dead.

DOO WAH DITTY

There I was just walking down the street,
Singing Doo Wah Diddy, Diddy, Dum Diddy Doo.

Recruiter walked up and started talking to me,
Singing Doo Wah Diddy, Diddy, Dum Diddy Doo.

It sounded good, the pay was fine,
When you get to boot camp,
You nearly lose your mind.

The next thing you know,
You're doing push ups CC,
Singing Doo Wah Diddy, Diddy, Dum Diddy Doo.

Now why don't you sound off,
One, two, three, four.
One, two, ...three-four.

MIGHTY SOLDIERS

We are Mighty soldiers,
Doing our country proud.
For duty, honor, country,
As much as we're allowed.

We are the best,
At doing what we do.
If you ever have a need,
We'll do our best to serve you.

We walk tall and proud,
As all soldiers should.
We'll fight for our country,
As you know that we would.

Left, right, left, right,
Your left, right, left, right.
We are Mighty soldiers,
Serving with all our might.

JODY BOY

Hey, Jody Boy,
Rough, tough, Jody Boy.
Hey, Jody Boy,
Rough, tough, Jody Boy.

I used to wear my faded jeans,
Now I'm wearing Army green.
I used to wear my faded jeans,
Now I'm wearing Army green.

Oh, Jody Boy,
Rough, tough, Jody Boy.
Oh, Jody Boy,
Rough, tough, Jody Boy.

I used to drive a pick up truck,
Now I pack it in my ruck,
I used to drive a pick up truck,
Now I pack it in my ruck.

Hey Jody Boy,
Rough, Tough,
Mean and Lean,
Jody Boy.

AIR FORCE BLUE

Air Force, Air Force, we come through,
We fly on by laughing at you.

Because we wear our blue like the sky,
The thing we do best 'course is fly.

HERE WE GO AGAIN

Here we go again,
Same old stuff again.
Marching down the avenue,
Ten more years and we'll be through.

Am I right or wrong?
(TROOPS) You're right!
Are we weak or strong?
(TROOPS) We're strong!

Sound off.
(TROOPS) 1, 2.
Sound off.
(TROOPS) 3, 4.
Bring it on down now.
(TROOPS) 1, 2, 3, 4.

MARINE PRIDE

You can keep your Army khaki and your Navy blues,
There's a different kind of fighting man I'll introduce you to.

His uniform is different,
The finest ever seen,
The Iraqis called him Devil Dog,
His real name is Marine.

He was born on Parris Island, The land that God forgot,
Where the sand is deep and the sun is blazing hot.

He'd walk a hundred miles before the day is done,
When it comes to fighting men,
The Marines are number one.

THEY SAY THAT IN THE ARMY

They say that in the Army,
The coffee is mighty fine.
Looks like muddy water,
And tastes like turpentine.

Oh Mom, I wanna go,
But they won't let me go,
Hoooooooooome, hey!

They say that in the Army,
The chicken is mighty fine.
One jumped off the table,
And started marking time.

Oh Mom, I wanna go,
But they won't let me go,
Hoooooooooome, hey!

They say that in the Army,
The rolls are mighty fine.
One rolled off the table,
And killed a friend of mine.

Oh Mom, I wanna go,
Drill Sergeant won't let me go,
Hoooooooooome, hey!

They say that in the Army,
The pay is mighty fine.
Give you one hundred dollars,
And take back ninety-nine.

Oh Mom, I wanna go,
But they won't let me go,
Hoooooooooome, hey!

MAMA, MAMA

Mama, mama, can't you see,
What the Navy's done to me.
Used to drive a Subaru,
Now I'm wearing Navy Blues.

Used to drive a Cadillac,
Pretty woman in the back.
Cut my hair and shaved my beard,
Now I'm looking mighty weird.
Misery, oh misery.
That's what the Navy is to me.

MARINE RECRUIT

Recruit, Recruit, can't you see,
Being mean is killing me.
Drill Instructor, all I see,
Is that you sure don't look sad to me.

Recruit, Recruit, I sure feel bad,
Walking and acting like I'm mad.
Drill Instructor, all I see,
Is that it doesn't look like an act to me.

Recruit, Recruit, where have you been,
Around this island and back again.
What ya gonna do when you get back,
I'm gonna take a shower and hit the rack.

DOWN BY THE RIVER

Down by the river,
We took a little walk.
Ran into China,
And had a little talk.

We pushed 'em
(TROOPS) Hey!
We pushed 'em
(TROOPS) Hey!

We threw 'em in the river,
And laughed while they drowned.
We don't need no Commies,
Hanging round.

(TROOPS)

Hey, don't be a fool,
Somebody said we we're number two.
They lied!
We're number ooooooooone, Army!

Down by the river,
We took a little walk.

Ran into Iraq,
And had a little talk.

We pushed ‘em
(TROOPS) Hey!
We pushed ‘em
(TROOPS) Hey!

We threw ‘em in the river,
And laughed while they drowned.
We don’t need no Saddam,
Hanging round.

(TROOPS)

Hey, don’t be a fool,
Somebody said we we’re number two.
They lied!
We’re number oooooooone, Army!

AIR FORCE BASIC

Lieutenant, Lieutenant, don't be blue,
Six more weeks and you'll be through.
When you get where you're going, you will know,
The Air Force is the way to go.

Look up, Look up in the sky,
F-15s are flying by.
Joined the Air Force to wear the blue,
So I can fly a fighter jet too.

IN THE EARLY MORNING RAIN

In the early morning rain,
In the early morning rain,
In the early morning raaaain,
In the early morning rain.

Dress it right and cover down,
Forty inches all around.
Dress it right and cover down,
Forty inches all around.

In the early morning rain,
In the early morning rain,
In the early morning raaaain,
In the early morning rain.

Sitting at home with nothing to do,
Got no money, no place to go.
Sitting at home with nothing to do,
Got no money, no place to go.

In the early morning rain,
In the early morning rain,
In the early morning raaaain,
In the early morning rain.

SEND ME

It was early in the morning,
Rounds were flying through the air.
Bombs were going off,
I was beginning to get scared.

I was in my fighting position,
My battle buddy was at my side.
The sergeant was not far away,
Ready to fight until we die.

As the sun hit the horizon,
I looked up to the beautiful sky.
And there I saw the greatest thing,
The Echo Company War Eagle Flying high.

I thought back to basic training,
And remembered everyone there.
From privates to the commander,
Suddenly I felt as if I were there.

It helped to calm me down,
To think of the people in those platoons.
And everyone that helped us then,
Especially the cadre that pulled us through.

Then in the midst of the battle,
I remembered a phrase dear to me.
And without any hesitation at all,
I said, “Send me Sergeant, send me”.

AIRBORNE RANGER

Airborne Ranger, Airborne Ranger,
Where have you been?
Around the world and back again.

Airborne Ranger, Airborne Ranger,
How did you go?
C-one thirty flying low.

Airborne Ranger, Airborne Ranger,
How'd you get down?
With a wide open parachute on my back.

Airborne Ranger, Airborne Ranger,
How'd you get back?
I humped one hundred miles with an old ruck sack.

AIRBORNE RANGER

Walking down the street one day,
I met a total stranger.
He asked me what I wanted to be,
I said an Airborne Ranger.

Airboooooooorrrrne,
Rangerrrrrrrrrrrrrr.

Sitting in a foxhole,
Sharpening my knife.
Up jumped the enemy,
I had to take his life.

Airboooooooorrrrne,
Rangerrrrrrrrrrrrrr.

I hear the choppers hovering,
They're hovering overhead.
They've come to get the wounded,
They've come to get the dead.

Airboooooooorrrrne,
Rangerrrrrrrrrrrrrr.

I went to wake the sergeant,
Lying in his bed.
When I rolled him over,
I saw that he was dead.

Airbooooooorrrrne,
Rangerrrrrrrrrrrrrrr.

THE ARMY LIFE

Keep your head and eyes off the ground,
Dress it right and cover down.
Marine and Airmen they're all right,
But we are better trained to flight.

Now be in peace or be in war,
We'll be here forevermore.
Wives and girlfriends are all alone,
Just waiting for the day that we get home.

Our top sergeant big and bad,
He treats us just like he was dad.
Mess hall chow is sure no treat,
But if you're hungry, you'll sure eat.

Working hard for the next stripe,
Can't get promoted 'til the time is right.
Give me life or give me death,
I'll be a soldier until my last breath.

Hear me once and hear me twice,
Reup now that's my advice.
We don't fear love and war,
We just run Jody out the door.

In columns left and columns right,
Dream of Drill Sergeants every night.
Love for moms and apple pie,
Makes me say I want to cry.

SCHOFIELD

Jody don't live in this here place,
He can't stand the heavy pace.

Back on the rock,
Marching up the block.

Even though the sky is blue,
We work too hard and this is true.

Back on the rock,
Marching up the block.

Hula girls and big palm trees,
You can't see them on K.P.

Back on the rock,
Marching up the block.

Prices are high and cash is low,
Savings don't get a chance to grow.

Back on the rock,
Marching up the block.

But in all, I like it here,
Magnum does, that's real clear.

Back on the rock,
Marching up the block.

LOOK OUT YOUR WINDOW

Hey, commander,
Look out your window.
Here comes your company,
Your well trained company.

We're marching home, hey!

Hey First Sergeant,
Look out your window.
Here comes your company,
You're rough, tough, company.

We're marching home, hey!

Sergeant Bishop,
Look out your window.
Here comes your platoon,
You're high speed platoon.

We're marching home, hey!

EARLY IN THE MORNING

Hi Ho, lock and load,
The engines rolling, ready to roll.
Kill the enemy, kill them slow.
So early in the morning.

Howitzer on the hill,
They raise the tube, they send the pill,
Send the pill, I know they will.
So early in the morning.

Bradleys creeping low,
The infantry is on the go,
Kill the enemy, I know they will.
So early in the morning.

Apache zooming high,
Hellfires in the sky,
Kill the enemy, I know I will.
So early in the morning.

MARINE CORPS SPIRIT

It was good at Pearl Harbor,
It was good at Iwo Jima.
It was good at Okinawa,
And it's good enough for me.

Give me that old Marine Corps spirit,
Give me that old Marine Corps spirit,
Give me that old Marine Corps spirit,
It's good enough for me.

It was good at Pusan Perimeter,
It was good at Frozen Chosen.
It was good in Korea,
And it's good enough for me.

Give me that old Marine Corps spirit,
Give me that old Marine Corps spirit,
Give me that old Marine Corps spirit,
It's good enough for me.

It was good at Koh Tang Island,
It was good for old Dan Dally,
It was good for Chessy Puller,
And It's good enough for me.

Give me that old Marine Corps spirit,
Give me that old Marine Corps spirit,
Give me that old Marine Corps spirit,
It's good enough for me.

It was good for Manila John Basilone,
It was good for General Lejeune,
It was good for Uncle Joe Pendleton,
And it's good enough for me.

Give me that old Marine Corps spirit,
Give me that old Marine Corps spirit,
Give me that old Marine Corps spirit,
It's good enough for me.

TINY BUBBLES

Tiny bubbles,
In my beer,
Makes me happy,
And full of cheer.

Your left, your left,
Your left, right, left.
Your left, your left,
Your left, right, left.

Tiny bubbles,
In my wine,
Makes me happy,
All the time.

Your left, your left,
Your left, right, left.
Your left, your left,
Your left, right, left.

Tiny bubbles,
In my milk,
Makes me happy,
Makes me wanna puke.

Your left, your left,
Your left, right, left.
Your left, your left,
Your left, right, left.

I DON'T KNOW WHY I LEFT

I don't know why I left,
But I know that I was wrong.
And it won't be long,
Until I get on back home.

Your left, your right, your left, your right,
Your left, your right, your out of sight.
And it won't be long,
Until I, until I, until I, get on back home.

Put me in the barber chair,
Turned around I had no hair.
And it won't be long,
Until I get back home.

Your left, your right, your left, your right,
Your left, your right, your out of sight.
And it won't be long,
Until I, until I, until I, get on back home.
Ain't no use in looking down,
Ain't no discharge on the ground.
Ain't no use in looking down,
Ain't no discharge on the ground.

Your left, your right, your left, your right,
Your left, your right, your out of sight.
And it won't be long,
Until I, until I, until I, get on back home.

MARINE CADENCE

Marine Cadence, Delayed Cadence, Count Cadence, Count.

"U"
I can't hear you.

"S"
A little louder.

"M"
Motivation.

"C"
Dedication.

KILL

Blood on the mountain top,
Blood on the sand.
The blood spells victory,
For our sweet land.

Eyes are focused,
With cammo on our face.
The enemy gets closer,
That's when we aim.

Shoot, shoot, shoot,
We're going to kill.
Kill, kill, kill,
With cold blue steel.

We are Echo Company,
We're gonna kill!

VICTORY

(Victory is a marching cadence that can be used with running cadence songs.)

V-I-C-T-O-R-Y.
Victory, Victory, that's my battle cry.

S-O-L-D-I-E-R
Soldiers, Soldiers, that's who we are.

C-O-U-N-T-R-Y
For duty, honor, country, I'd give my life.

C-one thirty rolling down the strip,
Airborne Ranger on a one way trip.
Stand up, hoop up, shuffle to the door,
Jump right out and count to four.

Mission undetermined, destination unknown.
We don't even know if we're ever coming home.

If my main don't open wide,
I got a reserve by my side.

If that one should fail me too,
Look out ground I'm coming through.

Bury me in the combat zone,
With my medals pinned on my chest.
Tell my mother not to cry,
And tell my father I did my best.

NAVY BLUES

Hup, two, three, four,
S and S then out the door.

Hup, two, three, four,
Hit the desks, mop the floors.

Hup, two, three, four,
Jog twice around, then twice more.

Hup, two, three, four,
Dad was right, it's a bore.

Hup, two, three, four,
All I'm hoping is a night on shore.

Hup, two, three, four,
Off the coast just waiting for war.

Hup, two, three, four,
Wishing for the mail a little more.

Hup, two, three, four,
So this is called a six month tour.

BACK IN THE DAY

Back in the day when I had no hope,
Back in the day when I had no trust.
Mama said go out and see the world,
So I packed my bags and got on a bus.

And it's gonna be a long day,
And it's gonna be a long night.
But I'm training to defend and fight.

Back in the day when I was scared to move,
Back in the day when I had to drop.
Drill Sergeant said one day you'll see the world,
So I pushed and pushed 'til he said stop.

And it's gonna be a long day,
And it's gonna be a long night.
But I'm training to defend and fight.

Back in the day when I became a soldier,
Back in the day when I showed some pride.
First Sergeant said one day you'll see the world,

So I swore from that day that I would never hide.

And it's gonna be a long day,
And it's gonna be a long night.
But I'm training to defend and fight.

One day soon I'll be an MP,
One day soon I'll graduate.
Captain said one day I'll see the world,
And I know that is my fate.

And I had long days,
And I had long nights,
But I trained to defend and fight.

HONEY, OH BABY

You get a line and I'll get a pole,
(TROOPS) Honey, Honey.
We'll go down to the crawdad hole,
(TROOPS) Babe, Babe.

(TROOPS)
You get a line, I'll get a pole,
We'll go down to the crawdad hole,
Honey, oh baby of mine.
Give me your left, your right, your left,
Give me your left, your right, your left, hey!

Dress it right and cover down,
(TROOPS) Honey, Honey.
Forty inches all around,
(TROOPS) Babe, Babe.

(TROOPS) Dress it right and cover down,
Forty inches all around,
Honey, oh baby of mine.
Give me your left, your right, your left,
Give me your left, your right, your left, hey!

I don't know but it's been said,

(TROOPS) Honey, Honey.
Communism must be dead,
(TROOPS) Babe, Babe.

(TROOPS)
I don't know but it's been said,
Communism must be dead,
Honey, oh baby of mine.
Give me your left, your right, your left,
Give me your left, your right, your left, hey!

PATCH ON MY SHOULDER

One O one,
Patch on my shoulder,
Screaming Eagles,
Pick up your weapon and follow me,
We're the Air Assault Infantry.

Eighty Second,
Patch on my shoulder.
Pick up your weapon and follow me,
We are the infantry.

MP LEAD THE WAY

Marching down the battlefield,
Look, listen, what's that I hear?
It was a simple recon mission,
That we were called out on.

But as the hills came closer,
I prayed for courage to stay strong.
Then I looked to my right,
To share my battle buddy's fright.

The air now filled with smoke,
As we dropped down to the prone.
And as we fought with all our heart,
I screamed with all my might.

No matter if it's night or day,
MP lead the way.

ARTILLERY, KING OF BATTLE

Artillery,
King of Battle.
We fought in World War One,
And we lead the way.

Artillery,
King of Battle.
We fought in Korea,
And we lead the way.

Artillery,
King of Battle.
We fought in Vietnam,
And we lead the way.

Artillery,
King of Battle.
We fought in Grenada,
And we lead the way.

Artillery,
King of Battle.
We fought in Panama,
And we lead the way.

Artillery,
King of Battle.
We fought in Desert Storm,
And we lead the way.

GRANDDADDY'S DITTY

My granddaddy was a Horse Marine,
When he was born, he was wearing green.

Ate his steak six-inches thick,
Picked his teeth with a swagger stick.

Drinking and fighting and running all day,
Granddaddy knew no other way.

Lived every day of his life for the Corps,
So they sent him off to war.

Went to the islands to fight the Japanese,
Caught some shrapnel in the knees.

Later, at Chosen Reservoir,
Caught a bullet in his derior.

Went to a country called Vietnam,
To fight some people called the Viet Cong.

Found himself in a fire-fight,
Came back home on a Medivac flight.
Now Granddaddy just sits there,

Marking time in his rocking chair.

FIRST IN, LAST OUT

Mighty U.S. Army,
Who holds your honor?
Who holds your pride?

Mighty U.S. Army,
Who will we send to battle?
Who will make it out alive?

MP's! First in, last out!
MP's! Will win, without a doubt!

Mighty Commander,
Who will you send?
Who will fight until the end?

Mighty Commander,
Who would do us proud?
Who would never fail?

MP's! First in, last out!
MP's! Will win, without a doubt!

Mighty First Sergeant,
Who would hold our flag high?

Who would never let it fall?

Mighty First Sergeant,
Who would die for us?
Who would win for us?

MP's! First in, last out!
MP's! Will win, without a doubt.

Mighty U.S. Army,
Mighty Commander,
Mighty First Sergeant,
Who holds your honor?
Who holds your pride?

MP's! First in, last out!
MP's! Will win, without a doubt.

OLD KING COLE

Old King Cole was a merry old soul,
And a merry old soul was he.

He called for his wife,
And he called for his pipe,
And he called for his privates three.

Beer, beer, beer, said the private.

Merry men are we,
There is none so fair
That they can compare
To the U.S. Army.

Old King Cole was a merry old soul,
And a merry old soul was he.

He called for his wife,
And he called for his pipe,
And he called for his corporals three.

I want a three day pass said the corporal.
Beer, beer, beer, said the private.
Merry men are we,

There is none so fair
That they can compare
To the U.S. Army.

Old King Cole was a merry old soul,
And a merry old soul was he.

He called for his wife,
And he called for his pipe,
And he called for his sergeants three.

I lead the way said the sergeant.
I want a three day pass said the corporal.
Beer, beer, beer, said the private.

Merry men are we,
There is none so fair
That they can compare
To the U.S. Army.

Old King Cole was a merry old soul,
And a merry old soul was he.

He called for his wife,
And he called for his pipe,
And he called for his louies three.

I want to be in charge cried the louie.
I lead the way said the sergeant.
I want a three day pass said the corporal.
Beer, beer, beer, said the private.

Merry men are we,
There is none so fair
That they can compare
To the U.S. Army.

SITTING AT HOME

Sitting at home, talking to Joe,
I've got no money, no place to go.
My recruiter came to me,
Said, what you want to be.

I told him mean, lean, and green,
An Airborne MP fighting machine.
So what can you do for me.

He took me down to the barber shop,
All my hair he made me chop.
My recruiter said to me,
Hey, what you want to be.

MP SOLDIER

I've put in many miles,
I've still got a lot to go.
The sun still hasn't risen,
The sky begins to glow.

M.Peeeeeeeeee
Soldierrrrrrrrrrr

My ruck sack's getting heavier,
With the things I need for life.
The ache runs through my body,
I continue through the strife.

M.Peeeeeeeeee
Soldierrrrrrrrrrr

I lead the way to battle,
I am the last to leave.
I try to tell my family,
Stories they won't believe.

M.Peeeeeeeeee
Soldierrrrrrrrrrr
We claim another victory,

For the red, white, and blue.
We'd fight another battle,
To keep the peace for me and you.

M.Peeeeeeeeee
Soldierrrrrrrrrrr

'TIL I GET ON BACK HOME

Your left, your right, your left, your right,
Your left, your right, you're out of sight.
And it won't be long,
'Til I get on back home.

Got a letter in the mail,
Said go to war or go to jail.
And it won't be long,
'Til I get on back home.

Your left, your right, your left, your right,
Your left, your right, you're out of sight.
And it won't be long,
'Til I get on back home.

Dress it right and cover down,
Forty inches all around.
And it won't be long,
'Til I get on back home.

Your left, your right, your left, your right,
Your left, your right, your out of sight.
And it won't be long,
'Til I get on back on home.

KNUCKLEHEAD

Oh, you knucklehead,
Dumb, Dumb, knucklehead.
Marching down the avenue,
Six more weeks and we'll be through.

Am I right or wrong?
(TROOPS) You're right.
Are we weak or strong?
(TROOPS) We're strong.

Oh, you knucklehead,
Dumb, dumb, knucklehead.
I'll be glad and so will you,
When all this training here is through.

Am I right or wrong?
(TROOPS) You're right.
Are we weak or strong?
(TROOPS) We're strong.

Oh you knucklehead,
Dumb, dumb, knucklehead.
Going to A.I.T.,
There's more in store for you and me.

Am I right or wrong?
(TROOPS) You're right.
Are we weak or strong?
(TROOPS) We're strong

MP, MP

We are MP's ready to stay,
Fighting for justice and the American way,
Red, white, and blue leading today.

MP, MP, brave and true,
Fighting for justice, striding for truth.

Leading the way in battle,
With a 9 millimeter held in aim,
MP, MP, lead the way.

Lead our Army to Victory,
We will die for our nation,
Forever our song will be sung.

YABBA, DABBA, DOO

Pebbles and Bam Bam on a Friday night,
Trying to get to heaven on a paper kite.

Singing yabba, dabba, dabba, doo,
Yabba, dabba, dabba, dabba, dabba, doo.

The lightning struck and down they fell,
Instead of going to heaven they went
Straight to hell.

Singing yabba, dabba, dabba, doo,
Yabba, dabba, dabba, dabba, dabba, doo.

HEY, HEY SWEET THING

Hey, hey sweet thing,
How do you do?
Do you remember me, baby,
The way I remember you?

We used to go to school,
My favorite subject was you.
The only "A" that I made,
Was on the homework you gave.

Hey, hey sweet thing,
How do you do?
Do you remember me, baby,
The way I remember you?

We used to meet by the lake,
You'd make my poor heart ache.
And by the old moonlight,
You'd always hold me tight.

Hey, hey sweet thing,
How do you do?
Do you remember me, baby,
The way I remember you?

You used to live over there,
By the railroad tracks.
And every time that it rained,
You used to call my name.

Hey, hey sweet thing,
How do you do?
Do you remember me, baby,
The way I remember you?

We used to sit by the stream,
And how we'd always dream.
And when we met at night,
You always treated me right.

Hey, hey sweet thing,
How do you do?
Do you remember me, baby,
The way I remember you?

COLUMBUS

In fourteen hundred and ninety-two,
Columbus sailed the Atlantic.
For days and days and weeks and weeks,
He sailed the broad Atlantic.

He said the world was roundo,
He said it could be foundo.
Hypothetical, Calculating,
Son-of-a-gun Columbo.

THERE'S A DRILL SERGEANT THERE

Everywhere I go,
There's a drill Sergeant there.
Everywhere I go,
There's a drill Sergeant there.

Drill Sergeant.
Drill Sergeant.
Why don't you leave me alone?
Why don't you let me go home?

When I get out of bed,
There's a drill Sergeant there.
When I get out of bed,
There's a drill Sergeant there.

Drill Sergeant.
Drill Sergeant.
Why don't you leave me alone?
Why don't you let me go home?

When I do PT,
There's a drill Sergeant there.
When I do PT,
There's a drill Sergeant there.

Drill Sergeant.
Drill Sergeant.
Why don't you leave me alone?
Why don't you let me go home?

When I go to chow,
There's a drill Sergeant there.
When I do to chow,
There's a drill Sergeant there.

Drill Sergeant.
Drill Sergeant.
Why don't you leave me alone?
Why don't you let me go home?

When I go to bed,
There's a drill Sergeant there.
When I go to bed,
There's a drill Sergeant there.

Drill Sergeant.
Drill Sergeant.
Why don't you leave me alone?
Why don't you let me go home?

HI, HO, DIDDLY BOP

Hi, Ho, diddly bop,
Wish I was back on the block.
With my sixteen in my hand,
I want to be a fighting man.
Fighting, fighting, all I can.

Your left, your left,
Your left, right, left.
Your military left,
Your left, your right,
Now pick up the step,
Your left, your right, your left.

Hi, ho, diddly bop,
Wish I was back on the block.
With my car keys in my hand,
I'm gonna be a driving man.
Driving, driving, all I can.

Your left, your left,
Your left, right, left.
Your military left,
Your left, your right,
Now pick up the step,

Your left, your right, your left.

Hi, ho, diddly bop,
Wish I was back on the block.
With my lover in my hands,
I'm gonna be a loving man.
Loving, loving, all I can.

Your left, your left,
Your left, right, left.
Your military left,
Your left, your right,
Now pick up the step,
Your left, your right, your left.

Hi, ho, diddly bop,
Wish I was back on the block.
With my boom-box in my hand,
I'm gonna be a rapping man.
Rapping, Rapping, all I can.

Your left, your left,
Your left, right, left.
Your military left,
Your left, your right,
Now pick up the step,

Your left, your right, your left.

IN SHAPE

Round the post and round we go,
Where we stop only Top knows.

Just for fun we pack 30 pound packs,
Bouncing real hard, bruising our backs.

For the Stars and Stripes we will run,
We'll do P.T. just for fun.

WE ARE SOLDIERS

Sergeant Major, he's a soldier,
He's got his hand on his leadership tab.
Then one day he couldn't lead anymore,
But got up and led anyhow.

(TROOPS)
Oh, oh, oh, we are soldiers, in the Army,
We gotta fight, oh we gotta fight, oh, oh, oh,
We gotta hold up the blood stained banner,
We gotta hold it up until we die.

Sergeant Timney, he's a soldier,
He's got his hand on his Airborne tab.
Then one day he couldn't jump anymore,
But got up and jumped anyhow.

(TROOPS)
Oh, oh, oh, we are soldiers, in the Army,
We gotta fight, oh we gotta fight, oh, oh, oh,
We gotta hold up the blood stained banner,
We gotta hold it up until we die.

Sergeant Williams, she's a soldier,
She's got her hand on her air assault tab.

Then one day she couldn't rappel anymore,
And got up and rappelled anyhow.

(TROOPS)
Oh, oh, oh, we are soldiers, in the Army,
We gotta fight, oh we gotta fight, oh, oh, oh,
We gotta hold up the blood stained banner,
We gotta hold it up until we die.

ROLL ALL OVER YOU

I'm a steam roller baby,
Just a rolling down the line.

I'm a steam roller baby,
Just a rolling down the line.

So you better get out of my way now,
Before I roll all over you.

With a little, a little, a little rock and roll.
With just the kind that, the kind that, the kind that soothes the soul.

So you better get out of my way now,
Before I roll all over you.

I'm an MP Trooper,
Just a rolling down the line.

I'm an MP Trooper,
Just a rolling down the line.

So you better get out of my way now,
Before I roll all over you.

With a little, a little, a little rock and roll.
With just the kind that, the kind that, the kind that soothes the soul.

So you better get out of my way now,
Before I jump all over you.

I'm an Airborne Ranger,
And I'm rolling down the line.

I'm an Airborne Ranger,
And I'm rolling down the line.

So you better get out of my way now,
Before I jump all over you.

With a little, a little, a little rock and roll.
With just the kind that, the kind that, the kind that soothes the soul.

So you better get out of my way now,
Before I jump all over you.

Singing, sha na na na na,
Sha, na na na na na na
Singing, sha na na na na

Sha, na na na na na na.

So you better get out of my way now,
Before I jump all over you.

MAMA, CAN'T YOU SEE

Mama, Mama, can't you see,
What the Army's done to me.

Mama, Mama, can't you see,
What the Army's done to me.

Put me on a silver jet,
The ground was cold, the air was wet.

Put me on a silver jet,
The ground was cold, the air was wet.

Mama, Mama, can't you see,
What the Army's done to me.

Mama, Mama, can't you see,
What the Army's done to me.

Took away my faded jeans,
Now I'm wearing Army green.

Took away my faded jeans,
Now I'm wearing Army green.
Mama, Mama, can't you see,

What the Army's done to me.

Mama, Mama, can't you see,
What the Army's done to me.

POP GOES THE WEASEL

All around the SAM site,
The missile chased the Weasel.

The Weasel got missed,
The SAM got zapped,
Pop goes the weasel.

Lady fighters did their jobs,
And more than just to tease them.

The Iraqi techs got all upset,
Pop goes the Weasel.

IT'S ALL RIGHT

It's all right, it's all right,
It's all right, it's OK.
It's all right, it's all right,
It's all right, it's OK.

It feels good, it feels good,
It feels good, it's OK.
It feels good, it feels good,
It feels good, it's OK.

Hey, hey, hey, hey,
Hey hey, hey, hey, hey,
It's gonna be OK.

It's all right, it's all right,
It's all right, it's all right,
It's all right, it's OK.

It's all right, it's alright,
It's all right, it's OK

SHA NA NA NA, SHA NA NA NA

Sha na na na, sha na na na,
Hey, hey, hey,we're going home.

No more Drill Sergeants,
No more Drill Sergeants,
Hey, hey, hey, we're going home.

No more chow hall,
No more chow hall,
Hey, hey, hey, we're going home.

Shan a na na, sha na na na,
Hey, hey, hey, we're going home.

Gonna graduate,
Gonna graduate,
Hey, hey, hey, we're going home.

Sha na na na, sha na na na,
Hey, hey, hey, we're going home.

UP ON YOUR LEFT

Up on your left, up on your left,
Up on your left, your right, your left,
Your right, left.

Ain't no use in looking down,
Your left, your right, left.

Ain't no discharge on the ground,
Your left, your right, left.

Singing, left, right,
All night.
Left, right,
All night.

(TROOPS)
Gee Mom, I wanna go 1, 2, 3, 4.
Up on your left, up on your left,
Up on your left, up on your left,
Your right, left, Hey!

I don't know but I've been told,
Your left, your right, your left.
First Platoon is mighty bold,

Your left, your right, your left.

Singing, left, right,
All night.
Left, right,
All night.

(TROOPS)
Gee Mom, I wanna go 1, 2, 3, 4.
Up on your left, up on your left,
Up on your left, up on your left,
Your right, left, Hey!

O HAIL, O HAIL MP

O hail, O hail MP,
If you're a Commando follow me.

O'hail, O hail MP,
If you're a Commando follow me.

Into the fight with an iron will,
Commandos are the first to kill.

Blood, guts, rifle, and knife,
Commandos will take your life.

Send us to war night or day,
Let a Commando lead the way.

Send us to war night or day,
Let a Commando lead the way.

We don't care if it's far or near,
Because commandos have no fear.

We don't care if it's far or near,
Because commandos have no fear.
Send a Commando and you will see,

That Commandos guarantee victory.

Send a Commando and you will see,
That Commandos guarantee victory.

We know someday, we will die.
But we'll never forget our battle cry.

Into the fight with an iron will,
Commandos are the first to kill.

Blood, guts, rifle, and knife,
Commandos will take your life.

PARATROOPER, PARATROOPER

Paratrooper, Paratrooper, where have you been?
Around the world and back again.

Paratrooper, Paratrooper, how'd you get down?
With a dash-one Bravo, big and round.

Paratrooper, Paratrooper, how did you fall?
With a rear parachute landing fall.

Paratrooper, Paratrooper, with what did you kill?
With Claymore Mines on that hill.

Paratrooper, Paratrooper, how'd you get back?
In a black and gold body sack.

Paratrooper, Paratrooper, how did you die?
With a seven point six two in the eye.

O HAIL, O HAIL MP (AIRBORNE)

O hail, O hail MP,
Airborne MP follow me.

Got a letter in the mail,
Go to war or go to jail.

O hail, O hail MP,
Airborne MP follow me.

Tell your mother not to cry,
Her little boy ain't gonna die.

O hail, O hail MP,
Airborne MP follow me.

They put me on a silver jet,
The air was cold, the ground was wet.

O hail, O hail MP,
Airborne MP follow me.

When the plane hit the ground.
There were Drill Sergeants all around.
O hail, O hail MP,

Airborne MP follow me.

They put me in a barber chair,
They spun me around, I had no hair.

O hail, O hail MP,
Airborne MP follow me.

I've seen the guts, I've seen the blood,
A lifeless body in the mud.

O hail, O hail MP,
Airborne MP follow me.

First Sergeant sends you off to chow,
You don't eat it anyhow.

O hail, O hail MP,
Airborne MP follow me.

Up in the morning and off of the rack,
Greeted by a mortar attack.

O hail, O hail MP,
Airborne MP follow me.

A long, long, day lies ahead.

O hail, O hail MP,
Airborne MP follow me.

Jungle fatigues, maroon beret,
Airborne MP leads the way.

O hail, O hail MP,
Airborne MP follow me.

Class “A”s and maroon beret,
Airborne MP tales a knee.
He says my darling marry me,

For overseas I must go,
I might not make it back you know.

O hail, O hail MP,
Lord of Death follow me.

Jungle fatigues, maroon beret,
Airborne MP leads the way.

Creeping through the jungle at night,
He is looking to take another life.

In fatigues and K-bar knife.
An MP takes another life.

He slits his throat from left to right,
Watches the blood run off the knife.

MP lying in the mud,
Covered in his own blood.

He begs you to stop the misery,
You lock and load your M-16.

A rifle shot rings out at night,
An MP loses his life.

He let out such an awful scream,
Wished that this was all a dream.

MP wife, she gets the check,
You look in her eyes as she opens it.

The moral is plain to see,
Nothing in this world is free.

O hail, O hail MP,
Airborne MP follow me.

DRIVE ON

Drive on First Platoon,
Some day you'll be alone,
Way out there in the combat zone.

Bullets flying all around,
Keep your head low to the ground.

Don't worry First Platoon,
Second Platoon will bring you home.

First squad is old and fat,
Ate up like a football bat.

Third Squad runs too low,
Drill Sergeant sounds off with 2-5-0.

HAIL O, HAIL O, FOLLOW ME

Thirteen men on recon,
They really thought they had it made.
Stepped out into the open,
All were killed by one grenade.

Hail O, Hail O, Infantry,
Queen of Battle follow me.
Hail O, Hail O, Infantry,
Pick up your weapon and follow me.

See the Commie on the hill,
Drop him dead for a thrill.
See the rag head in the sand,
Blow him away where he stands.

Hail O, Hail O, Infantry,
Queen of Battle follow me.
Hail O, Hail O, Infantry,
Pick up your weapon and follow me.

ASSIST THE BATTLE

I hear you calling,
Calling for me.
Assist the battle,
Airborne MP.

I've fought in trenches of mud,
I've fought in trenches of blood.

I hear you calling,
Calling for me.
Assist the battle,
Airborne MP.

I have been covered in mud,
I have been covered in blood.

I hear you calling,
Calling for me.
Assist the battle,
Airborne MP.

I have been chest high in mud,
I have been chest high in blood.
I hear you calling,

Calling for me.
Assist the battle,
Airborne MP.

I have been cold and wet,
I have been covered in sweat.

I hear you calling,
Calling for me.
Assist the battle,
Airborne MP.

UGLY CADENCE, COUNT

Count Cadence,
Delay Cadence,
Count Cadence,
COUNT!

(TROOPS) “U”

You are ugly.

(TROOPS) “G”

And your brother too.

(TROOPS) “L”

I am so glad.

(TROOPS) “Y”

I don’t look like you.

(TROOPS) “U”

Hit it!

(TROOPS) “G”

Hit it!

(TROOPS) “L”

Hit it!

(TROOPS) “Y”

Hit it!

(TROOPS)
U-G-L-Y You ain’t got no alibi,
You’re ugly, hey, hey, you’re ugly.

LEFT, RIGHT, NOW GET ON DOWN

Your left, right, your left,
Your left, right, now get on down.

Used to drive a Chevy truck now,
Now I pack it in my ruck now.

Your left, right, your left,
Your left, right, now get on down.

Dress it right and cover down now,
Forty inches all around now.

Your left, right, your left,
Your left, right, now get on down.

Mama, Mama, can't you see now,
What the Army's done to me now.

Your left, right, your left,
Your left, right, now get on down.

ALL THROUGH THE NIGHT

One by one,
We're having some fun in the Army,
All day and all through the night, hey, hey.

Two by two,
Just me and you killing Commies,
All day and all through the night, hey, hey.

Three by three,
Just you and me going airborne,
All day and all through the night, hey, hey.

Four by four,
We'll kill us some more,
Killing commies,
All day and all through the night, hey, hey.

Five by five,
We'll stand them in line,
Killing Commies,
All day and all through the night, hey, hey.

Six by six,
We'll do it for kicks,

Killing commies,
All day and all through the night, hey, hey.

Seven by seven,
They won't go to heaven,
Not the commies,
All day and all through the night, hey, hey.

Eight by eight,
We're feeling great,
In the Army,
All day and all through the night.

Nine by nine,
We're marking time,
In the Army,
All day and all through the night, hey, hey.

Ten by ten
We'd do it again,
Killing Commies,
All day and all through the night, hey, hey.

MP COUNT CADENCE

Count Cadence, Delay Cadence, Count Cadence, COUNT!

(TROOPS) One!

MP SOLDIER

(TROOPS) Two!

Better do your best.

(TROOPS) Three!

Before you find yourself.

(TROOPS) Four!

In the leaning rest.

(TROOPS) One!

Hit it!

(TROOPS) Two!

Hit it!

(TROOPS) Three!

Hit it!

(TROOPS) Four!

(TROOPS)
One, two, three, four, Military Police Corps,
M-I-L-I-T-A-R-Y-P-O-L-I-C-E!

LOOKING GOOD

Standing tall and looking good,
We ought to be in Hollywood.
Standing tall and looking good,
We ought to be in Hollywood.

Dress it right and cover down,
Forty inches all around.
Dress it right and cover down,
Forty inches all around.

Nine in front and six to the rear,
That's the way we do it here.
Nine in the front and six to the rear,
That's the way we do it here.

Hold your head and hold it high,
Second Platoon is marching by.
Hold your head and hold it high,
Second Platoon is marching by.

ARE MADE FOR

These boots are made for walking,
That's what they're gonna do.
If all your doing is marking time,
They'll walk all over you.

The Army's trained for fighting,
That's what we're gonna do.
If you try to mess with us,
We'll run all over you.

Guns are made for shooting,
That's what they're gonna do.
If we catch you in the woodline,
We'll drill holes all over you.

YELLOW RIBBON

Around her hair she wore a yellow ribbon,
She wore it in the Spring time
And in the merry month of May.

And if you asked her why the heck she wore it,
She told you for her soldier who was far, far, away.

Far away,
Far away,
She wore it for her soldier who was far, far, away.

Around the block she pushed a baby carriage,
She pushed it in the Spring time
And the merry month of May.

And if you asked her why the heck she wore it,
She told you for her soldier who was far, far, away.
Far away,

Far away,
She wore it for her soldier who was far, far, away.

Behind the door her daddy kept a shotgun,
He kept it in the Spring time and the merry moth of May.

And if you asked him why the heck he kept it,
He told you for her soldier who was far, far, away.

Far away,
Far away,
She wore it for her soldier who was far, far, away.

SEE THAT MAN

See that man with a Green Beret,
Killing's how he earns his pay.
That's the only life for me,
Specials Forces Infantry.

See that man with a Maroon Beret,
Jumping how he earns his pay.
That's the only life for me,
Drill Sergeant Infantry.

See that man with the Round Brown on,
Training troops all day long.
That's the only life for me,
Drill Sergeant Infantry.

See that man with the Tan Beret,
Humpin's how he earns his pay.
That's the only life for me,
Ranger, Ranger, Infantry.

See that man with the Soft Cap on,
Marching, marching, all day long.
That's the only life for me,
Trainee, Trainee, Trainee Infantry.

THE ARMY COLORS

The Army colors,
The color is red.
To show the world,
The blood we shed.

That Army colors,
The color is blue.
To show the world,
That we are true.

The Army colors,
The color is green.
To show the world,
We're lean and mean.

The Army colors,
The color is white.
To show the world,
We're fit to fight.

The Army colors,
The color is brown.
To show the world,
When we're in town.

The Air Force color,
The color is blue.
To show the world,
That they're true too.

MY GIRL

My girls is a pretty girl,
She's a New York city girl.
I'd buy her anything,
To keep her in style.

She has a head of hair,
Just like a grizzly bear.
I'd give her anything,
To keep her in style.

She has a pair of eyes,
Looks like two custard pies.
I'd buy her anything,
To keep her in style.

She has a long, long, nose,
Just like a garden hose.
I'd buy her anything,
To keep her in style.

She has a pointy chin,
Just like a safety pin.
I'd buy her anything,
To keep her in style.

She has a pair of hips,
Just like two battle ships.
I'd buy her anything,
To keep her in style.

She has a pair of knees,
Just like the summer breeze.
I'd buy her anything,
To keep her in style.

She has a pair of feet,
Just like a parakeet.
I'd buy her anything,
To keep her in style.

RANGERS ARE MOVING

Rangers are moving,
They're moving through the night.
The Rangers are moving,
They're looking for a fight.

And if you see one,
A Ranger standing tall.
If you listen well,
You hear the Ranger call.

Airborrrrrrrrrrrrrne,
Rangerrrrrrrrrrrrr.

When Iraqi's moving,
Moving through the night.
Iraqi, Iraqi,
He thinks he can fight.

When up from behind him,
A Ranger standing tall.
He cut the Iraqi's throat,
And watched the Iraqi fall.

Airborrrrrrrrrrrrrne,

Rangerrrrrrrrrrrrr.

MEN AT WAR

Sitting in a fox hole,
Sharpening my knife.
When all of the sudden,
My buddy yelled grenade.

Men at waaaaaarrrrr, kill,
Men at waaaaaarrrrr, kill.
Late at night when you're sleeping,
A Drill Sergeant's always creeping all around,
Men at waaaaaarrrrr, kill.

Sitting on a hill top,
Eating beans and franks.
When all of the sudden,
The bullets started back.

Men at waaaaaarrrrr, kill,
Men at waaaaaarrrrr, kill.
Late at night when you're sleeping,
A Drill Sergeant's always creeping all around,
Men at waaaaaarrrrr, kill.

SOUND OFF!

Herby, Herby had big feet,
Stretched from here to Market Street.
Market Street is made of glass,
Herby fell and cut his elboq.

SOUND OFF!

(TROOPS) One, Two.

SOUND OFF!

(TROOPS) Three, four,

Bring it on down now.

(TROOPS)
One, two, three, four, one, two....three, four.

Soldier, Soldier on the wall,
Ain't you got no sense at all.
Can't you see that wall's been plastered,
Get off that wall, you silly little spider.

SOUND OFF!

(TROOPS) One, Two.

SOUND OFF!

(TROOPS) Three, four,

Bring it on down now.

(TROOPS)
One, two, three, four, one, two….three, four.

JUMP FROM A BIG OLD BIRD

Airborne, Airborne, have you heard,
We're gonna jump from a big ole bird.

Stand up, hook up, shuffle to the door,
Jump right out and count two, three, four.

If my main don't open wide,
I got a reserve by my side.

If that one should fail me too,
I'm a ground dart, pure and true.

Scoop me up and send me home,
Back to the land where the bad legs roam.

Now I'll be back, mark this day,
Airborne, Airborne, all the way.

MODERN MILITARY CADENCE®
CURRENT OPERATIONS

RUNNING CADENCES

SITTING ON A MOUNTAIN TOP

Sitting on a mountain beating a drum,
Beat so hard 'till the MP's come.

MP, MP, don't arrest me,
Arrest that Grunt behind the tree.

He stole the whisky, he stole the wine,
All I do is double time.

Cause I'm hard core,
Mean and lean,
Cut clean,
Looking good,
Hollywood,
Every day,
One mile,
No sweat,
Two miles,
Better yet.

One, two, three, four, hey.
One, two, three, four, hey.

JUMP, SWIM, KILL

I saw an old lady running down the street,
She had a chute on her back and boots on her feet.

I said old lady where you going to,
She said the U.S. Army Airborne School.

I said old you're too darn old,
Leave that training to the young and the bold.

I saw an old man coming down the track,
He had fins on his feet and tanks on his back.

I said old man where are you going to,
He said U.S. Army Scuba School.

I saw a young man coming down the road,
He had a knife in his hand and a ninety pound load.
I said young man where you going to,
He said U.S. Army Ranger School.

What you gonna learn when you get there?
How to jump, swim, and kill without a care.

CAPTAIN JACK

Hey, hey, Captain Jack,
Meet me down by the railroad tracks.

With that weapon in my hands,
I'm gonna be a killing man.
Killing, killing, all I can.

Hey, hey, Captain Jack,
Meet me down by the railroad tracks.

With that knife in my hands,
I'm gonna be a cutting man.
Cutting, cutting all I can.

Hey, hey, Captain Jack,
Meet me down by the railroad tracks.

With those car keys in my hand,
I'm gonna be a driving man.
Driving, driving, all I can.

Hey, hey, Captain Jack,
Meet me down by the railroad tracks.
I'm gonna giver her all my charm.

Charm her, charm her all I can.

Hey, hey, Captain Jack,
Meet me down by the railroad tracks.

MP LEAD THE WAY

Four thirty in the morning,
MP's on the move.
With a mission to accomplish,
And something to prove.

With a two mile run,
We start the day.
Army MP's
Lead the way.

I said MP,
Soldier of choice,
You can see it in their eyes,
You can hear it in their voice.

I said MP,
Charging straight ahead,
Taking care of the living,
Bringing home the dead.

From Bosnia to Saudi,
Everywhere in between .
MP's are there looking,
Lean, mean and green.

Protect and serve,
By our motto we live.
Our lives for our Country,
We truly give.

I said MP,
Soldier of Choice,
You can see it in their eyes,
You can hear it in their voice.

I said MP,
Charging straight ahead,
Taking care of the living,
Bringing home the dead.

IN THE MOVIES

Movie stars have nothing on us,
Our moves are made without that fuss.

Panama was a practice run,
Jumping in Grenada was just for fun.

Rambo could learn a thing or two,
We left him behind, nursing the flu.

Cameras ready, crew on call,
We're marching by and standing tall.

KEEP IT IN STEP

Left, left, left, your righta, left.
Left, left, keep it in step.

Oh righta left,
Keep it in step.

Oh righta, left.
We love to double time.

We do it all the time.

AMERICAN WAY

Working hard for Uncle Sam,
Ready to fight for my fellow man.

Freedom, freedom, that's what I say,
Fighting for the American way.

Forever we hold our banner high,
We'll hold it up forever until we die.

Winning wars is what we do,
Fighting hard for me and you.

HEY, HEY, YOU CAN'T STOP

Here comes Rangers, over the hill,
Rolling like a wagon wheel.

Hey, hey, you can't stop the mean and green,
Hey, hey, you can't stop a fighting machine.

Here come Rangers 'round the bend,
Doing what they do again.

Hey, hey, you can't stop the mean and green,
Hey, hey, you can't stop a fighting machine. \

Airborne Rangers over the hill,
Rolling like a wagon wheel.

Hey, hey, you can't stop the mean and green,
Hey, hey, you can't stop a fighting machine.

RUNNING PAPA

Momma and Papa were lying in bed,
Papa rolled over to Mama and said.

Give me some P.T.,
It's good for you,
It's good for me.

We get up in the morning with the rising sun,
We run all day 'till the day is done.

Mile one, having fun,
Mile two, good for you.

Mile three, good for me,
Mile four, I want some more.

Mile five, keeps me alive,
Mile six, great kicks.

Mile seven, this is heaven,
Mile eight, this is great.

Mile nine, mighty fine,
Mile ten, let's do it again.

C-130 ROLLING

C-130 rolling down the strip,
Airborne MP on a one way trip.

Stand up, hook up, shuffle to the door,
Jump right out and count to four.

Mission undetermined, destination unknown,
We don't even know if we're ever coming home.

If my main don't open wide,
I got ma reserve beside my side.

If that one should fail me too,
Look out ground, I'm coming through.

Pin my medals upon my chest,
Bury me in the leaning rest.

Tell my mother not to cry,
Tell my father I did my best.

OH ONE, TWO, THREE, FOUR

Oh one, two, three, four,
Drop and beat your face some more,
Oh one, two, three, four,
Military Police Corps.

Oh one, two, three, four,
Push ups, sit ups, three mile run,
Oh one, two, three, four,
U.S. Army number one.

Oh one, two, three, four,
Your left, your right, you're out of sight,
Oh one, two, three, four,
Get up, get down, side straddle hop.

Oh one, two, three, four,
High speed soldiers on their way,,
Oh one, two, three, four,
You're the Army of the day.

Oh one, two, three, four,
Left, right, on left, right,
Oh one, two, three, four,
MP soldiers coming through.

Oh one, two, three, four,
Stop look we're better than you,
Oh one, two, three, four,
Dress right and cover down.

Oh one, two, three, four,
Echo Company's number one.

WE ARE AIRBORNE, YEA

We are Airborne, yea,
Mighty, Mighty, Ranger, yea.

We can fly higher than Superman.

We are Airborne, yea,
Mighty, Mighty, Ranger, yea.

We are stronger than all your men, yea.

We are Airborne, yea,
Mighty, Mighty Ranger, yea.

We can swim faster than Aquaman.

We are Airborne, yea,
Mighty, Mighty Ranger, yea.

We can climb higher than Spiderman.

We are Airborne, yea,
Mighty, Mighty Ranger, yea.

STREET SMART

The Crips and Bloods could learn from us,
Stabbing and shooting, ready to cuss.

We do what we're trained to do,
Fighting for freedom for me and you.

Street gangs stealing and shooting up drugs,
We know what to teach those thugs.

Use those muscles on Saddam Hussein,
Show them street smarts is the name of the game.

RUNNING 'ROUND

Running 'round (post name) and we'll never stop.
Keep on running and don't dare stop.

Running 'round (post name) and we're having fun,
We kill Commies one by one.

THE MP CORPS

Run me, run me some more,
I'm a soldier in the MP corps.

When I got there it was shark attack,
Drill Sergeants were all on my back.

Push-ups, sit ups, two mile run,
Now we do P.T. for fun.

First there was B.R.M.,
If you messed up, you shot again.

Then there was the disco hut,
Where all us privates puked our guts.

Now we're in A.I.T.,
Were twice a day we do P.T.

Classes for you and classes for me,
Classes on how to be an MP.

Next thing you know, we graduate,
Then we'll work shifts from 8 to 8.
You want to be like me, Hard Core, MP.

ARMY VS. MARINES

I don't know, but I've been told,
The Marine Corps thinks it's mighty bold.

They don't know what the Army can do,
We are proud of our history too.

Our looks and style may not be smooth,
But you ought to see this Army move.

Look to your left and what do you see?
A bunch of Jarheads just looking at me.

Shout it out and sing it loud,
I'm a soldier and I'm mighty proud.

DOG TAG DANGLE

One, two, three, four, hey.
One, two, three, four, hey.

Don't let your dog tags dangle in the dirt,
Pick up your dog tags and tuck'em in your shirt.

Don't let your dog tags get into a bend,
Don't lend your dog tags to a buddy or a friend.

One, two, three, four, hey.
One, two, three, four, hey.

HI, HO DIDDLY BOP

Hi, ho, diddly bop,
I wish I was back on the block.

With that bottle in my hand,
I'm gonna be a drinking man,
Drinking soda all I can.

Your left, your left,
Your left, right, left.
Your Third Platoon left,
Pick up the step, your left, right, left.

Hi, ho, diddly bop,
I wish I was back on the block.

With that book on the shelf,
I'm gonna educate myself.
Reading, studying, all I can.

Your left, your left,
Your left, right, left.
Your Third Platoon left,
Pick up the step, your left, right, left.
Hi ho, diddly bop,

I wish I was back on the block.

With that remote in my hand,
I'm gonna be a T.V. man,
Watching T.V. all I can.

Your left, your left,
Your left, right, left.
Your Third Platoon left,
Pick up the step, your left, right, left.

Hi ho, diddly bop,
I wish I was back on the block.

TAN BERET

Hey, all the way,
We run every day.

Airborne Rangers with the Tan Beret,
We're Kamikaze killers and we earn our pay.

Jumping out of airplanes,
Running through swamp,
Uncle Sam gets in trouble,
Rangers gonna stomp.

Our minds are like computers,
Our fists are like steel,
If one don't get you,
The other one will.

We do push ups every morning,
Our bodies are rocks,
We'd run to Alabama,
But we'd wear out our socks.

WHEN I GET TO HEAVEN

When I get to heaven,
Saint Peter's gonna say.

How'd you earn your living?
How'd you earn your pay?

And I'll reply with a little bit of anger,
Made my living as an Airborne Ranger.

Airborne Ranger,
Ranger Danger.
Airborne Ranger,
Tan Beret danger.

I love to double time,
I do it all the time.

When I get to heaven,
Saint Peter's gonna say.

How'd you earn your living?
How'd you earn your pay?

And I'll reply with a little fit of anger,

Made my living blood, guts and danger.

Blood, guts and danger,
Airborne Ranger.
Airborne Ranger,
Blood, guts, and danger.

I love to double time,
I do it all the time.

MICHAEL JACKSON

Michael Jackson came to town,
Coca Cola turned him down.

Pepsi Cola burned him up,
Now he's drinking Seven Up.

GOING TO IRAQ

I'm going over to Iraq,
With a rifle and a ruck on my back.

When I'm done kicking Saddam's butt,
I'm going down to Columbia.

We're gonna move the druggies out,
We're gonna move them without a doubt.

Kicking and fighting and cutting all day,
We don't know no other way.

MP SOLID AS A ROCK

MP, solid as a rock hoo-a,
Wish I was back on the block.

With that law book in my hand,
I'm gonna be a law abiding man,
Abiding the law all I can.

One, two, three, four, hey,
Run me, run me, run me some more, hey.

MP, solid as a rock hoo-a,
Wish I was back on the block.

With that ticket book in my hands,
I'm gonna be a writing man,
Writing tickets all I can.

One, two, three, four, hey,
Run me, run me, run me some more, hey.

MP solid as a rock hoo-a,
Wish I was back on the block.

RECRUITING RANGERS

My recruiter- he told me a lie,
He said join the Rangers and learn to fly.

He said sign my name on the dotted line,
Now all I'm doing is double time.

Hey, all the way,
We run every day.

Hey, all the way,
We run every day.

YOUR SON

Mothers of American, meek and mild,
Send to me your sweet young child.

We'll make him drill and make him run,
We'll make some changes in your son.

Mothers of America, meek and mild,
Say goodbye to your sweet child.

We're gonna make him drill,
We're gonna make him run,
We're gonna make some changes in your son.

MP'S

We're MP's see how we roar,
Clean up the street, then beg for more.

Throw out your crack and dump your beer,
Better look sharp, MP's are near.

Use your signal, stop at signs,
Don't drive too fast, just toe the line.

Fight with your spouse and we'll be there,
When you do something wrong, better
beware!

IN STEP

Left, left, left, your righta, left,
Left, left, keep it in step.

Oh righta, left,
Keep it in step.

Oh righta left,
I love to double time,
I do it all the time.

START A WAR

One, two, three, four, hey,
Somebody, anybody, start a war, hey.

A Leer Jet's watching,
On a midnight run,
With an M-16,
And an Ingrim gun.

We halo in,
We stable out.
Death from above,
You can hear us shout.

With gun in hand,
Where the days are long.
From Bamba land,
To Panama.

Three young men,
In a Russian truck.
With a little Mac-ten,
Seen'em run into the hut.

These young men,

The ones who dare.
Do battle in hell,
The mercenary.

UP IN THE MORNING

Up in the morning at the break of day,
Working so hard, we never play.

Running in jungles where the sun don't shine,
All I do is double time.

Up in the morning and out of the rack,
Grab my clothes and put'em on my back.

Running 'cross the desert in the shifting sand,
Drill Sergeant, look, I need a helping hand.

Hey, hey, Ft. Benning you can hide your face,
Because Ft. (fill in post name)'s gonna breeze by you.

DADDY WENT AWAY

Well, I was just a baby when my daddy went away,
Killed in action or so they say.

Living up the good life on a Mississippi ferry,
I'm gonna go to Columbia and be a mercenary.

Gonna even that battle, gonna even that score, Gonna kill me a Commie with a mine, Claymore.

JUMP BOOTS

Tooled skin, and alligator hide,
Make a pair of jump boots just the right size.

Don't be afraid to put them on your feet, A good pair of jump boots can't be beat.

Shine 'em up, lace 'em up, put 'em on your feet,
Brand new jump boots can't be beat.

C-130 RANGER

C-130 rolling down the strip,
Sixty-four Rangers on a one-way trip.

Mission unspoken, destination unknown,
They don't even know if they're ever coming home.

Stand up, hook up, shuffle to the door,
Jump right out and count to four.

If that main don't open wide,
You got a reserve by your side.

And if that one should fail you too,
Look out below 'cause I'm coming through.

I hit the ground in the middle of the night,
Spring right into a firefight.

PAPPY BOYINGTON

Pappy Boyington ruled the air,
The Japanese thought he fought unfair.

Twenty-eight Zeros he shot down,
Pappy was the best fighter pilot around.

A medal of Honor received with grace,
Pappy was the Marine Corps' Number One Ace.

SEEN AN OLD LADY

Seen an old lady walking down the street,
Had a ruck on her back and jump boots on her feet.

Said, old lady, where are you going to?
She said, the U.S. Army Airborne School.

I said, hey, old lady, ain't you been told,
Airborne School is for the young and the bold.

Hey, young man, who you talking to,
I'm an instructor at the Airborne School.

Seen an old lady walking down the street,
Had a ruck on her back and jump boots on her feet.

Said, old lady, where are you going to?
She said, the U.S. Army Ranger School.

I said, hey, old lady, ain't you been told,
Ranger school is for the young and the bold.

Hey, young man, who you talking to,

I'm an instructor at the Ranger School.

Seen an old lady walking down the street,
She had tanks on her back and fins on her feet.

Said, old lady, where you going to?
She said, U.S. Army Diving School.

I said, hey old lady, ain't you been told,
Diving school is for the strong and the bold.

Hey, young man, who you talking to,
I'm the lead instructor at the Diving School.

AIRBORNE SOLDIER

Soldier, soldier, have you heard,
I'm gonna jump from a big iron bird.

Up in the morning in the drizzling rain,
Packed my 'chute and boarded the plane.

Raining so hard that I couldn't see,
Jumpmaster said, "Depend on me."

I looked with fear at the open door,
Then I stood up and fainted on the floor.

When I woke up, I was hooked up again,
Then my eyes rolled back and I fainted again.

HIT THE BEACH

Up from a sub fifty feet below,
Up swims a man with a tab of gold.

Back stroke, side stroke, headed for shore,
He hits the beach, he's ready for war.

Grease gun, K-bar by his side,
These are the tools that he lives by.

How to kill a search team, a hostage snatch,
Out to the sub, back into the hatch.

Hand to hand combat behind the lines,
Airborne Ranger justa having a ball.

UP IN THE MORNING

Up in the morning, too soon,
Eat my breakfast before noon.

Went to the mess sergeant on my knees,
Mess sergeant, mess sergeant, feel me please.

Mess sergeant said with a big old grin,
If you want to be Infantry, you gotta be thin.

Up in the morning, too soon,
Eat my breakfast before noon.

Went to the mess sergeant on my knees,
Mess sergeant, mess sergeant, feel me please.

Mess sergeant said with a big old grin,
If you want to be Airborne, you gotta be thin.

If you want to be a Ranger, you gotta be thin.

DOUBLE TIME

Double time, double time, up the hill,
Everybody's gonna get a two mile thrill.

Double time, double time, everyone will,
Everybody's gonna get their fill.

Double time, double time, two miles long,
It's impossible to go wrong.

Double time, double time, to this song,
Everybody's gonna make it strong.

Double time, double time, going strong,
We didn't know we could run so long.

Double time, double time, all the way,
We can run like this everyday.

I love double time, you love it too,
We can run the whole day through.

Double time, double time, having fun,
We can't wait for the five mile run.

RANGER ROCK

Up in the morning in the pouring rain,
Donned my ‘chute and I boarded a plane.

One Thirty Herc rolling down the strop,
Sixty-four Rangers on a killing trip.

Mission top secretly, destination unknown,
Airborne Ranger ain’t coming home.

‘Cause he’s hard core,
Ranger,
Rock,
Steady,
Fit to fight,
Rock,
Steady,
Easy,
Easy.

Stand up, hook up, shuffle to the door,
My knees got weak and I fell to the floor.

The Jump Master picked me up with ease,
And threw my knees into the breeze.

'Cause he's hard core,
Ranger,
Rock,
Steady,
Fit to fight,
Rock,
Steady,
Easy,
Easy.

FOLLOW ME

Hey, hey, there Army,
Get in your tanks and follow me,
I'm the Marine Corps Infantry.

Hey, hey, there Navy,
Get in your ships and follow me,
I'm Marine Corps Infantry.

Hey, hey, there Air Force,
Get in your planes and follow me,
I'm Marine Corps Infantry.

DEALING DEATH

Up jumped a Ranger from the Seventy-fifth,
He had vengeance in his eyes he was dealing death.
He was a Recon Scout on a halo trip,
Made his money killing Commies didn't ever quit.

Flying over Moscow in the middle of the night,
Twenty five grand above the city lights.

He made it to the Kremlin,
He was creeping down the hall,
Heard'em talking politics and listened to it all.

Then kicked down the door and drove into the room,
He was the last thing the Russians saw
before they met their doom.

With a Carb Fifteen and a couple of Frags,
Now they run their country out of red body bags.

AIRBORNE DADDY

Papa told Sally not to go downtown,
All the Airborne were hanging around.

Sally went ahead and went on down,
Now she's back with her belly round.

Sally's Ma said to her what ya gonna do,
We just can't take care of you.

Sally told her Ma don't worry 'bout me,
My Airborne can take care of three.

MILITARY POLICE

MP,MP, what you gonna do?
Catch that crook, follow that clue.

MP, MP, where you gonna go?
Ticket that cycle for going slow.

MP, MP, how have you been?
Can't complain, I just can't sin.

MP,MP, why you holding that gun?
I'm just about to have some fun.

MP,MP, did you see that fight?
She hit him with a left and then with her right.

MP,MP, have you heard the news?
SGT Jones has the short timers blues.

MP,MP, why you gonna ticket me?
'Cause the party was wild and the booze was free.

MP,MP, when do you plan to sleep?
No, been up all night, I can't count sheep.

WE ARE RANGERS

Johnny, you're a bad boy, sad boy, bad boy,
Come to Ranger School and he ain't gonna
Make it; Got the cammy on his face, a big
disgrace, Got the ruck sack all over the place.

(TROOPS)
We are, we are, Rangers (CLAP, CLAP)
Rangers,
We are, we are, Rangers (CLAP, CLAP)
Rangers.

Johnny, you're a bad boy, sad boy, bad boy,
You go to third bat and he ain't gonna make
it,
Got the cammy on his face, a big disgrace,
Got the ruck sack all over the place.

(TROOPS)
We are, we are, Rangers (CLAP, CLAP)
Rangers,
We are, we are, Rangers (CLAP, CLAP)
Rangers.

Johnny, you're a bad boy, sad boy, bad boy,

Go back to Ranger School and he ain't gonna Make it; Got the cammy on his face, a big disgrace, Got the ruck sack all over the place.

(TROOPS)
We are, we are, Rangers (CLAP, CLAP)
Rangers,
We are, we are, Rangers (CLAP, CLAP)
Rangers.

ROLLIN', ROLLIN', ROLLIN'

Rollin', rollin', rollin',
Oh, my head is swollen.

Don't let your dog tags dangle in the dirt,
Pick up your dog tags and tuck 'em in your shirt.

Rollin', rollin', rollin',
Oh, my neck is swollen.

Don't let your dog tags dangle in the dirt,
Pick up your dog tags and tuck 'em in your shirt.

Rollin', rollin', rollin',
Oh, my cheek is swollen.

Don't let your dog tags dangle in the dirt,
Pick up your dog tags and tuck 'em in your shirt.

Rollin', rollin', rollin',
Oh, my hips are swollen.
Don't let your dog tags dangle in the dirt,

Pick up your dog tags and tuck 'em in your shirt.

Rollin', rollin', rollin',
Oh my thighs are swollen.

Don't let your dog tags dangle in the dirt,
Pick up your dog tags and tuck 'em in your shirt.

Rollin', rollin', rollin',
Oh my calves are swollen.

Don't let your dog tags dangle in the dirt,
Pick up your dog tags and tuck 'em in your shirt.

Rollin', rollin', rollin',
Oh my feet are swollen.

Don't let your dog tags dangle in the dirt,
Pick up your dog tags and tuck 'em in your shirt.

AIRBORNE (SUNG TO AMEN)

Here we go now.
(TROOPS) Airborne.

All together now.
(TROOPS) Airborne.

One more time now,
(TROOPS) Airborne, Airborne, Airborne.

Here we go now,
(TROOPS) Airborne.

Feels so good now,
(TROOPS) Airborne.

All together now,
(TROOPS) Airborne, Airborne, Airborne.

OLD LADY DIVER

Saw an old lady walking down the street,
She had tanks on her back and fins on her feet.

Hey, old lady, where you going to?
United States Navy Diving School.

Hey old lady ain't you been told,
Scuba diving's for the young and the bold.

Sonny, sonny, can't you see,
I'm an instructor at U.D.T.

DRILL INSTRUCTOR

I want to be a Drill Instructor,
I want to cut off all of my hair.

I want to be a Drill Instructor,
I want to earn that Smokey Bear.

MANILA JOHN BASILONE

John Basilone was a tough man,
Carried a machine gun in each hand.

Cut japs down with the machine gun fire,
As he dodged bullets in the quagmire.

A Medal of Honor was on his chest,
Proved he was better than the rest.

MARINE RECON RANGER

Up from a sub fifty feet below,
Swam to the surface and I'm ready to go.

I want to be a Recon Ranger,
I want to live a life of danger.
Recon Ranger,
Life of danger.

Grease gun, K-Bar by my side,
These are the tools that make men die.

I want to be a Recon Ranger,
I want to live a life of danger.
Recon Ranger,
Life of danger.

Side stroke, back stroke, swim to shore,
Hit the beach and I'm ready for war.

I want to be a Recon Ranger,
I want to live a life of danger.
Recon Ranger,
Life of danger.

MISTER PT

Mister PT is playing my song,
Running fast and marching all day long.

Mister PT is singing my note,
Cut it short and you'll get my vote.

Mister PT is singing my tune,
I want a hot shower, gonna hit it soon.

Mister PT is attempting to shine,
Marching all the way up the Rhine.

Mister PT is learning to rap,
Gonna dance Karlsruhe on the map.

Mister PT is trying to roll,
He's even trying to run to Seoul.

Mister PT is making me see double,
We better quit before he's in trouble.

MARINE PARATROOPER

Tell my mother not to cry,
The Marine Corps life is do or die.

Place my K-Bar in my hand,
I'll fight my way to the promise land.

Hooking and a jabbing, sticking and a stabbing, cutting and a slicing, kicking and a fighting.

FLYING

Flying high and flying low,
Watching the Infantry on the go.

Flying in and flying out,
Hear the Infantry scream and shout.

Flying up and flying down,
I'll gig up the enemy shooting up the ground.

Flying forward and flying back,
I'll help the grunts in their attack.

SOLEMN VOW

Through the desert, across the plains,
Steaming jungles and tropic rains.

No mortal foe can stop me now,
This is gonna be my solemn vow.

I have honor and I have pride,
Winning serves me as my guide.

This Army shocks our enemies,
Bringing them crashing to their knees.

Basic training is plenty rough,
To make it through, you must be tough.

Squad leader, don't be blue,
They're gonna make you a soldier too.

CHESTY PULLER

Chesty Puller was tough and mean,
He was a one man fighting machine.

Chesty's heroics were a matter of lore,
He was the toughest fighter in the Corps.

Chesty conquered every test,
Five Navy Crosses on his chest.

HERE WE GO

Here we go,
All the way,
Every day,
PT,
Feeling good,
Looking good,
Ought to be,
Hollywood,
Dress it right,
Cover down,
Forty inches,
All around,
You can do it,
PT,
One mile,
No sweat,
Two miles,
Better yet,
Here we go,
All the way.

ARMY, NAVY

Army, Navy, where you been?
Down to P.I. and back again.

Army, Navy, what do you see?
D.I. students running PT.

Singing left, right, left.

Army, Navy, what did you do?
Will, I jumped right in and I PT'd too.

Singing left, right, left.

Army, Navy, what did you think?
Well, I think I want to be a U.S. Marine.

The sing left, right, left.

Air Force, Air Force, don't be blue,
The Marines are gonna teach you how to PT too.

We sing left, right, left.

AIR FORCE

F-15 rolling down the strip,
Eagle Driver gonna take a little trip.

Rev it, taxi up, count to four
Push the throttle forward and
Hear the engines roar.

ThirtyMy M thousand feet up in the air,
Flying this baby is a natural high.

Took a look at 6 o'clock and
What did I see,
A MIG-21 was coming after me.

Pulled it up, rolled it out,
Much to his surprise,
Should have seen the look,
In that turkey's eyes.

Got behind him, set my sights,
Let my missile fly,
Blew that 21 out of the sky.

When you see an Eagle Driver,

He will say,
Flying and fighting is the Air Force Way.

BURGER KING

Down in Honolulu at the Burger King,
First Sergeant Jones was a doing his thing.

Hamburger, hot dog, chocolate shake,
There isn't much that he can't take.

Stand up, hustle up, shuffle to the door,
Back to the track and run some more.

MARINE CORPS COLORS

My Marine Corps color,
The color is green.
To show the world,
That we are mean.

My Marine Corps color,
The color is gold.
To show the world,
That we are bold.

My Marine Corps color,
The color is red.
To show the world,
The blood we shed.

My Marine Corps color,
The color is Black.
To show the world,
That we fight back.

My Marine Corps color,
The color is Blue.
To show the world,
That we are true.

AIRBORNE RANGER ALL THE WAY

Jesse James said before he died,
There's five things that he wanted to ride.

Bicycle, tricycle, automobile,
Bow legged horse and a ferris wheel.

'Cause he's Airborne Ranger all the way.

Down in the jungle where the coconuts grow,
There is a mean motor scooter named Ranger Joe.

He's in all the way up to his knees,
He's an Airborne Ranger coming through the trees.

'Cause he's Airborne Ranger all the way.

THIS SAILOR LOVES YOU

Baby, baby, don't be blue,
Because this sailor still loves you.

Even though I am far away,
You're in my thoughts every day.

On this ship or on this shore,
You're the one that I adore.

And no matter where I roam,
Baby you know I'm coming home.

So if someone questions you,
Tell them that our love is true.

And if you think that I know,
How if feels to be alone.

Even though I'm with the boys,
None of them replaces our joys.

As we sing this song,
Now you know where I belong.

I CAN RUN TO...

Up in the morning, half past three,
First Sergeant Fischer was bringing heat.

Had NCO's all around his desk,
Had the Company Commander in the leaning rest.

First Sergeant Fischer can't you see,
You can't bring no smoke on me.

I can run to Maryland just like this,
All the way to Baltimore and never quit.

I can run to Florida just like this,
All the way to Miami just like this.

I can run to Washington just like this,
All the way to D.C. and never quit.

I can run to California just like this,
All the way to L.A. and never quit.

I can run to Alabama just like this,
All the way to Anniston and never quit.

ARMY LIFE

Oh, I joined the Army ranks,
Just so I could see a tank.

Oh, I left a nagging wife,
Just to live this Army life.

Oh, this Army is for me,
It's the only place to be.

I like woman and I like wine,
But all I do is double time.

Double time here, double time there,
Army life is the best anywhere.

WANNA BE

If you wanna be a doggy,
Then you should have joined the Army.
Hey, hey, yea,
Hey, hey, yo.

If you wanna be a swabbie,
Then you should have joined the Navy.
Hey, hey, yea,
Hey, hey, yo.

If you wanna be a fly boy,
Then you should have joined the Air Force.
Hey, hey, yea,
Hey, hey, yo.

If you wanna be a mean and lean,
You gotta join the Marines.
Hey, hey, yea,
Hey, hey, yo.

If you want to be an officer,
Then you have to go to Quantico.
Hey, hey, yea,
Hey, hey, yo.

But, if you want to be a D.I.,
You have to go to San Diego.
Hey, hey, yea.
Hey, hey, yo.

EXTRA, EXTRA

Extra, Extra, hot from the press,
Navy saves the Army from a great big mess.

Extra, Extra, hear the news,
Navy comes through, Army is blue.

Extra, Extra, it ain't no jive,
Navy keeps those Army boys alive.

MARINE RECRUITER

I used to sit at home all day,
Letting my life waste away.

Then one day a man in blue said,
Son, I got the job for you.

Free room and board with pay to boot,
And a brand new tailored suit.

There's travel, adventure, and loads of fun,
You'll even learn to shoot a gun.

Free medical benefits, your body tight,
You get eight hours of sleep at night.

Believe me, lad, he said with a grin,
Just sign these papers and I'll get you in.

As I got on the bus, he winked his eye,
I never knew he could be so sly.

I'm not complaining, don't get me wrong,
I've enjoyed myself growing big and strong.
But two things he did not promise me,

Were a rose garden and a cup of tea.

SUPERMAN

I don't know, but I think I can,
Take the S from Superman.

Early in the morning, late last night,
Me and Superman had a fight.

I hit'em with my left, I hit'em with my right,
I hit'em in the heat with some kryptonite.

I hit'em so hard I busted his brain,
Now I'm dating Lois Lane.

Jimmy Olsen jumped me from the rear
I kicked'em in the knee and bit'em on the ear.

I kicked'em with my foot and hit'em with my hand,
Now he's buried with Superman.

BATMAN

I don't know, but I think I might,
Look for Batman Saturday night.

I don't know, but I think I will,
Take a little ride in the bat mobile.

Me and Batman we fought too,
I kicked'em in the head with the heel of my shoe.

I kicked'em with the toe, kicked'em with the heel.
Now I'm kicking in the bat mobile.

ALL THE RIGHT STUFF

Mamma, mamma, don't you cry,
Your baby little boy ain't gonna die.

I'll make him tough, I'll make him strong,
You won't know'em when he gets gone.

He'll be strong and he'll be tough,
He'll be made of all the right stuff.

Loyalty, duty, and respect,
A lot of selfless service you can bet.

And when he speaks he'll be heard,
Honor and integrity in his words.

Leadership it is the key,
It's the backbone of the Army.

ALMIGHTY INFANTRY

I am the almighty Infantry,
Pick up your weapons and follow me.

I hump all day and I hump all night,
On my way to the firefight.

We're trained to kill, we're trained to fight,
We do it all, we do it all night.

This is my family, this is my team,
U.S. Army fighting machine.

Together we'll conquer the enemy,
Cause we're the almighty infantry.

KEEP IT TIGHT

Mamma, mamma, don't you cry,
Your little boy ain't gonna die.

1 push up, 2 push ups, 3 push ups, 4,
My best friend is the concrete floor.

Drill Sergeant runs us all day and night,
His favorite line is ,"keep it tight!"

Keep it tight
Dress it right
Cover down

One hundred and ten is what we give,
That's the only way we're gonna live.

Cause I'm hard core,
Infantry.

RUNNING FOR MY HEALTH

Left, left, left, right oh left,
Your right left,
Stay in step,
Left, right, left,
Running for my health.

I love to double time,
I do it all the time,
It makes me feel fine,
Clears up my mind.

Left, left, left, right oh left,
Your right left,
Stay in step,
Left, right, left,
I'm running for my health.

DOG TAGS

Rollin', rollin', rollin',
Ohhhh my feet are swollen.

Don't let your dog tags dangle in the dirt,
Pick up your dog tags and tuck 'em in your shirt.

Rollin', rollin', rollin',
Ohhhh my knee is swollen.

Don't let your dog tags dangle in the rocks,
Pick up your dog tags and tuck 'em in your socks.

Rollin', rollin', rollin',
Ohhhh my back is swollen.

Don't let your dog tags dangle in the dirt,
Pick up your dog tags and tuck 'em in your hand.

Rollin', rollin', rollin',
Ohhhh my head is swollen.
Don't let your dog tags dangle in the dirt,

Pick up your dog tags and hand 'em to your bud.

ROLLIN' ROLLIN' ROLLIN'

Rollin', rollin', rollin',
Oh my _________ are (is) swollen.
(ankles, feet, legs, knees, etc...)

Don't let your dingle-dangle dangle in the mud.
Pick up your dingle-dangle, give it to your bud.

Don't let your dingle-dangle dangle in the dirt.
Pick up your dingle-dangle, put it in your shirt.

Don't let your dingle-dangle dangle on the ground.
Pick up your dingle-dangle, shake it all around.

Don't let your dingle-dangle dangle on the track.
Pick up your dingle-dangle, put it in your pack.

DINGLE-DANGLE

Left, left, oh right left,
Left, right, left
Keep it in step.

But, don't let your dingle-dangle dangle in the dirt,
Pick up your dingle-dangle, put it in your shirt.

Left, left, oh right left,
Left, right, left
Keep it in step.

But, don't let your dingle-dangle dangle in the mud.
Pick up your dingle-dangle, hand it to your bud.

Left, left, oh right left,
Left, right, left
Keep it in step.

But, don't let your dingle-dangle dangle in the snow,
Pick up your dingle-dangle, tie it in a bow.

Left, left, oh right left,
Left, right, left
Keep it in step.

But, don't let your dingle-dangle dangle too low.
Pick up your dingle-dangle, and let's go.

Left, left, oh right left,
Left, right, left
Keep it in step.

MY TWO DOGS

I got a dog, his name is Jack,
If you throw him a stick, he won't bring it back.

He's got better things to do,
You see, old Jack is Airborne too.

Jack be nimble, Jack be hard,
Not a dog for miles comes in my yard.

Jack be quick, Jack be bad,
Jack has got his Ranger tab.

When he was a little pup,
We made him a chute and took him up.

Standing in the door, there was no doubt,
He was wagging his tail when he went out.

Cause he's A.I.
R.B.
O.R.
N.E.
Airborne,
Ranger,

All the way.

I got another dog, I named him Blue,
Blue wanna be a Ranger too.

So, early one day I took away his chow,
And put some motivation in his bow wow.

I made him walk for 15 days,
And put old Blue into a zombie haze.

Now my Blue is a Ranger too,
Mess with him and he'll bite you.

Cause he's A.I.
R.B.
O.R.
N.E.
Airborne,
Ranger,
All the way.

BLUE

Had an old dog who's name was Blue,
Blue wants to go to SCUBA school.

Bought him a tank and four little fins,
And took him down where he got the bends.

Same old dog who's name was Blue,
He now wants to go to Ranger School.

Took him to the field, took away his chow,
Put a little motivation in his bow wow.

Still had the dog who's name was Blue,
He now wants to go to Airborne School.

Got him a chute, strapped it to his back,
Now old Blue stands tall, looks strack.

That Airborne dog who's name was Blue,
Got his orders for Jungle School.

Took him on down to Panama,
And tha's the last of Blue I ever saw.

ONE MILE

Here we go
On the run
Just for fun

One mile
I can hang

Two miles
You can hang

Three miles
We can hang

Ha!
Ha, ha
Hooah
Ha, ha

Four miles
I can do it

Five miles
We can do it
Six miles

Nothing to it

Ha!
Ha, ha
Hooah
Ha, ha

Little run
To the sun

Ha!
Ha, ha
Hooah
Ha, ha

I can hang
Can you hang

I can hang
With the pain

ONE MILE

One mile
No sweat

Two miles
Better yet

Three miles
Gotta run

Four miles
To the sun

HEY UP FRONT

Hey up front if you want to bump, say what

(Chorus) Don't stop the PT run
or Don't wash your PT shorts

Hey in the middle if you want to wiggle, say what

(Chorus) Don't stop the PT run
or Don't wash your PT shorts

Hey in the rear is if you want some beer, say what

(Chorus) Don't stop the PT run
or Don't wash your PT shorts

A little louder

(Chorus) Don't stop the PT run
or Don't wash your PT shorts

OLD LADY AIRBORNE

I saw an old lady running down the street,
Had a chute on her back, jump boots on her feet.

I said, hey old lady where you going to,
She said U.S. Army Airborne School.

What you gonna do when you get there,
Jump from a plane and fall through the air.

I said, hey old lady ain't you been told
Airborne school is for the brave and the bold.

She said, hey now soldier, don't be a fool,
I'm an instructor at the Airborne School.

OLD LADY RANGER

I saw an old lady marching down the road,
Had a knife in her hand and a 90 pound load.

I said hey old lady where you going to,
She said U.S. Army Ranger School.

What you gonna do when you get there,
Jump, swim, and kill without a care.

I said, hey old lady ain't you been told,
Ranger school is for the brave and the bold.

She said, hey young soldier don't be a fool,
I'm the lead instructor at Ranger School.

OLD LADY DIVER

I saw an old lady running 'round the track,
Had fins on her feet and a tank on her back.

I said, hey old lady where you going to,
She said, U.S. Army SCUBA school.

What you gonna do when you get there,
Swim under water and never breathe air.

I said, hey old lady ain't you been told,
SCUBA school is for the young and the bold.

She said, hey now diver don't be a fool,
I'm the head instructor at the SCUBA school.

UP IN THE MORNING

I don't know, but I think I might,
Jump from an airplane while in-flight.

Soldier, soldier have you heard,
I'm gonna jump from a big iron bird.

Up in the morning in the drizzling rain,
Packed my chute and boarded the plane.

C-130 rollin' down the strip,
Airborne Rangers on a one-way trip.

Mission top secret, destination unknown,
Don't even know if we're ever coming home.

When my plane gets up so high,
Airborne troopers gonna dance in the sky.

Stand up, hook up, shuffle to the door,
Jump right out and count to four.

If my main don't open wide,
I got a reserve by my side.
If that one should fail me to,

Look out ground I'm coming through.

If I die on the old drop zone,
Box me up and ship me home.

Bury speakers all around my head,
So I can rock with the Grateful Dead.

Bury speakers all around my toes,
So I can rock with Axel Rose.

If I die on a Chinese hill,
Take my watch or the commies will.

If I die in the Korean mud,
Bury me with a case of Bud.

Put my wings upon my chest,
And tell my mom I did my best.

UP AND OUT OF THE RACK

Up in the morning and out of the rack,
Grab my clothes and put them on my back.

Running ‘cross the desert in the shifting sand,
Drill sergeant look I’ll give you a helping hand.

Up in the morning with a whistle and a yell,
I know that voice and I know it well.

Drill sergeant says you better hit the floor
And don’t be walking going out the door.

I like fun and I like wine,
But all I do is double time.

Double time here and double time there,
Man this life, it’s the best anywhere.

GRANNY DOES PT

When my granny was 91,
She did PT just for fun.

When my granny was 92,
She did PT better than you.

When my granny was 93,
She did PT better than me.

When my granny was 94,
She did PT more and more.

When my granny was 95,
She did PT to stay alive.

When my granny was 96,
She did PT just for kicks.

When my granny was 97,
She upped, she died, and went to heaven.

When my granny was 98,
She did PT at the pearly gates.
When my granny was 99,

She was in heaven doing double time.

MY OLD GRANNY

When my old granny was 91,
She joined the Army just for fun.

When my old granny was 92,
She did PT in combat boots.

When my old granny was 93,
She practiced PLFs from a tree.

When my old granny was 94,
She knocked out 10 and begged for more.

When my old granny was 95,
She fired expert and that's no jive.

When my old granny was 96,
She went Airborne just for kicks.

When my old granny was 97,
She up and died and went to heaven.

She met St. Peter at the pearly gates,
Said hey St. Peter I hope I ain't late.
St. Peter looked at granny with a big ol' grin,

Said get down granny knock out ten.

She knocked out ten and did ten more,
Dedicated them to the NCO corps.

Peter looked at granny and said you're so cool,
We're sending you back for Ranger school.

Granny said to Peter, hey I ain't no fool,
I could be teaching at that dog gone school.

A.I.R.B.O.R.N.E.

A-is for Airborne

I-is for in the sky

R-is for ranger

B-is for Bonafied

O-is for on the go

N-is for never quit

E-is for everyday

Cause I'm Airborne all the way
Super-duper paratrooper.

RAVING MAD

Airborne Ranger raving mad,

He's got a tab I wish I had.

Black and gold and a half moon shape,

Airborne Ranger he's gone ape.

Jumpin' through windows, kicking down the walls,

Airborne Ranger's having a ball.

So if there's trouble in the world today,

Call on the men in the tan berets.

WHEN I GET TO HEAVEN

When I get to heaven,
Saint Peter's gonna say.

How'd you earn your living,
How'd you earn pay.

And I'll reply with a little bit of anger,
Made my living as an Airborne Ranger.

Airborne Ranger
Ranger danger
Airborne Ranger
Tan beret danger

I love to double time,
I do it all the time.

EARN YOUR LIVING

When I get to heaven,
St. Peter's gonna say.

How'd you earn your living boy,
How'd you earn your pay.

I'll reply with a whole lot of anger,
Made my living as an Airborne Ranger.

Blood and guts and a whole lot of danger,
That's the life of an Airborne Ranger.

When I get to hell,
Satan's gonna say.

How'd you earn your living boy,
How'd you earn your pay.

I'll reply with a fist to his face,
Made my living laying souls to waste.

C-130

C-130 rollin down the strip,
Airborne daddy on a one way trip.

Mission uncertain, destination unknown,
We don't know if we're ever coming home.

Stand up, hook up, shuffle to the door,
Jump right out and count to four.

If my main don't open wide,
I got a reserve by my side.

If that one should fail me too,
Look out ground I'm coming through.

Slip to the right, and slip to the left,
Slip on down, do a PLF.

Hit the drop zone with my feet apart,
Legs in my stomach, feet in my heart.

BURY ME

If I die on the old drop zone,
Box me up and ship me home.

Pin my wings upon my chest,
Bury me in the leaning rest.

If I die in the Spanish Moors
Bury me deep with a case of Coors.

Of I die in Korean Mud,
Bury me deep with a case of Bud.

If I die in a firefight,
Bury me deep with a case Lite.

If I die in a German Blitz,
Bury me deep with a case of Schlitz.

If I die, don't bring me back,
Just bury me with a case of Jack.

CHAIRBORNE RANGER

It's one thirty now on the strip,
Chairborne daddy gonna take a little trip.

Stand up, lock up, shuffle to the door,
The club for lunch and home by four.

If there's something to decide,
Close your door and try to hide.

Every time you get a call,
You're out playing raquetball.

First revise the SOP,
Make a change in policy.

Ours is not to wonder why,
It's written down in the LOI.

God forbid we got to war,
All that paperwork would be a bore.

Let me stay behind my desk,
Anything is better than the leaning rest.
Chairborne Ranger, that's what I am,

One of a kind, I'm an AG man.

COON SKIN

Coon skin and alligator hide,
Make a pair of jump boots just the right size.

Shine 'em up, lace 'em up, put 'em on your feet,
A good pair of jump boots can't be beat.

Birdy, birdy, in the sky,
Dropped some white wash in my eye.

Ain't no sissy, I won't cry,
I'm just glad that cows don't fly.

NOTHING TO DO

AG, AG, who are you?
TDA with nothing to do.

Go to PT at 9 am,
Then to the pool to have a swim.

Racquetball from 9 to 10,
Recover with a tonic and gin.

Lunch from 11 until noon,
You day will be over soon.

Volleyball from noon 'til 3,
Keep really busy, can't you see.

Now it's 4, your day is through,
I wish I was AG too.

AMEN

Aaaaaaaaaaaamen (chorus continuously by all).

(Caller sings between chorus lines)

Sing it over

Sing it louder now

Sing it soft now

Real loud now

Real soft now

Hallelujah

Praise the Lord now

G.I.

G.I. coat and G.I. comb,
Gee, I wish that I were home.

G.I. coat and G.I. gravy,
Gee, I wish I'd joined the Navy.

WHERE'VE YOU BEEN

AWOL, AWOL, where've you been
Down in the bar, drinking gin.

What you gonna do when you get back,
Sweat it all out on the PT track.

ENGINEERS ARE NUMBER ONE

Engineers are number one,
They call us when there's work to be done.

Aviation is all we hear,
We do the work, they drink the beer.

Pistol ranges, soccer goals,
Road extensions, we do it all.

ENGINEER, ENGINEER

Engineer, Engineer, running down the road,
Running so fast makes the others look old.

We are running hard we're running long,
Still singing another stupid song.

Build a road or cut down a tree,
Or dig some graves for the Infantry.

Working hard and working all day,
Knocking down anything that gets in the way.

DESERT SAND

Went down to see the man,
He gave me orders for the desert sand.

I packed my weapon, packed up my ruck,
They threw me in this 5 ton truck.

As I look out with a glassy glare,
The next thing I know I'm in the air.

When we land, it's dark and warm,
They tell me I'm at the Desert Storm.

For the next six months, this is your home,
No running water, no telephone.

Saddam Hussein he said to me,
I want to be all I can be.

I'll pack your weapon, I'll pack your ruck,
As for Iraq they have no luck.

FORCE RECON

Paint my face black and green,
You won't see me I'm a recon Marine.

I slip and slither into the night,
You won't see me 'til I'm ready to fight.

You'll run in the bushes, you'll try to hide,
But that's where I live, you're sure to die.

You won't see me 'til it's too late,
A flash of my bang will be your fate.

GEORGE S. PATTON

In 1934 we took a little trip,
Me and George S. Patton headed down to Mississipp'.

We shot our main guns 'til the barrels melted down,
Then we grabbed a couple legs and we went a couple rounds.

Cause we're mentally able and we're physically fit,
And if you ain't Armor, you ain't it.

GRANDADDY

My Granddaddy was a horse Marine,
When he was born, he was wearing green.

Ate his steak six inches thick,
Picked his teeth with a swagger stick.

Drinking and fighting and running all day,
Granddaddy knew no other way.

Lived everyday of his life for the corps,
So they sent him off to war.

Went to the islands to fight the Japanese,
Caught some shrapnel in the knees.

Later, at Chosen Reservoir,
Caught a bullet in his derriere.

Went to a country called Viet Nam,
To fight some people called the Viet Cong.

Found himself in a firefight,
Came back home on a Medevac Flight.
Now Granddaddy just sits there,

Marking his time in his rocking chair.

IRAQI BLUES

Send the troops before it's too late,
Saddam has invaded Kuwait.

Grab your rifle and get a tan,
You can scratch a rotation plan.

President Bush was talking tough,
We didn't know it would get that rough.

Thought Saddam was a man of reason,
Now we've got him for the crime of treason.

America's become divided as such,
They don't like that war stuff much.

Cussing and picketing that's the scoop,
Throw rocks at me but you support our
troops.

People are starting to understand,
Saddam Hussein's one crazy man.

Gasses his people and tortures them too,
Saddam this cluster bomb's for you.

Burning oil and acid rain,
SCUD missile, desert terrain.

Shipped my butt straight overseas,
Hey who cut down all the trees.

One, two, three, and four,
Sometimes to get peace, you gotta make some war.

If we don't nuke 'em 'til they glow,
We'll die for more than Texaco.

Stormin' Norman made a plan,
Now we're gonna kill who's in command.

When we're through kicking his butt,
We'll pay for gas and it won't cost much.

This is my story and it is true,
I call this song the Iraqi Blues.

Saddam, act stupid and I won't refuse,
To put you on the 10 o'clock news.

JUMP INTO BATTLE

Jump, jump, jump, into battle,
Hear those sixty's rattle.

Shoot, move, and cover my brother,
Write a letter to my mother.

Jump, jump, jump, into battle,
Hear those fifty cals rattle.

Shoot, move and cover my brother,
Write a letter to my mother.

MARINE BY GOD

Born in the woods, raised by a bear,
I got a double set of jaw teeth and a triple coat of hair.

Two brass balls and a cast iron rod,
I'm a mean devil dog, a Marine by God.

1775

Back in 1775,
My Marine Corps came alive.

First there came the color blue,
To show the world that we are true.

Next there came the color red,
To show the world the blood we shed.

Finally there came the color green,
To show the world that we are mean.

TALE OF THE RECON MARINE

Way, way back in the dawn of time,
In the valley of death where the sun don't shine.

A mighty fighting man was made,
From an M-16 and a live grenade.

He looked mighty big with his ALICE pack,
He drove mighty mean with his Cadillac.

This mighty fighting lean green machine,
Goes by the title, Recon Marine.

Roll on your left and roll on your right,
Roll on your left we love to double time.

TARZAN AND JANE

Tarzan and Jane were swinging from a vine,
Sippin' from a bottle of whisky double wine.

Jane missed the vine and then she fell,
When she hit the ground, she gave a little yell.

Ai-e-ai
Mmm mmm
Feels good
Ai-e-ai
Mmm mmm
Real good

Tarzan and Cheetah were swinging from a vine,
Sippin' from a bottle of whisky double wine.

Cheetah missed the vine and then he fell,
When he hit the ground, he gave a little yell.

Ai-e-ai
Mmm mmm
Feels good

Ai-e-ai
Mmm mmm
Real good

TERRIBLE JAM IN VIET NAM

Come on all you big strong men,
Uncle Sam need your help again.

Got himself in a terrible jam,
Way down yonder in Viet Nam.

So put down your books and pick up a gun,
We are all gonna have a whole lotta fun.

Come on Wall Street don't be slow,
Man this is war so go, go, go.

There is a lot of good money to be made,
Supplying the Army with the tools of the trade.

Just hope and pray that if we drop the bomb,
We go and drop it on Viet Nam.

Come on generals, let's move fast,
Your big chance is here at last.

Now we can go out and get those reds,

Cause the only good commie is one that's dead.

UP FROM A SUB

Out in the sky in the middle of the night,
When we hit the deck we're ready to fight.

Up from a sub 60 feet below,
We SCUBA to the surface and we're ready to go.

We're gonna back stroke, side stroke, swim to shore,
When we hit the beach, we're ready for war.

Singing hoo-ya, hoo-ya, hey
Hoo-ya, running day,
Singing hoo-ya, hoo-ya, hey,
Just another PT day.

Well chief caught a round right between the eyes,
And corpsman thought for sure the chief would die.

But chief stood up straight as any man,
And killed four commies hand-to-hand.
Well 20 seconds later there was not a sound,

And 50 dead commies were lying around.

Singing hoo-ya, hoo-ya, hey
Hoo-ya, running day,
Singing hoo-ya, hoo-ya, hey,
Just another PT day.

Now superman may be the man of steel,
But he ain't no match for a Navy Seal.

Now chief and superman they got in a fight,
Chief hit him in the head with some
kryptonite.

Superman fell down on his knees in pain,
Now chief is dating Lois Lane.

Singing hoo-ya, hoo-ya, hey
Hoo-ya, running day,
Singing hoo-ya, hoo-ya, hey,
Just another PT day.

SCUBA BLUE

Well I've got a dog and his name is Blue,
And Blue wants to be a Navy Seal too.

So, I bought him a mask and four little fins,
I took him to the ocean and threw his butt in.

Blue came back and to my surprise,
With a shark in his teeth and a gleam in his eyes.

WHEN I DIE

When I die, burry me deep,
With two crossed rifles laid beneath my feet.

By my side a forty five I wear,
And don't forget to pack my PT gear.

Because early one morning around zero five,
The ground's gonna shake, they'll be thunder in the sky.

Don't you get alarmed, don 't you come undone,
It's just me a Chesty Puller on a PT run.

FILLERS

Yea, I'm hard-core

Lean and mean

Fit to fight

Out of sight

One mile

No sweat

Two miles

Better yet

Three miles

I can make it

You can make it

Huah
A – ha

Huah

A – ha

Hard-core

Lean and mean

On the scene

Army green

UH-60

UH-60 flying high,
These are the wimps that snivel and cry.

Stand up, snap in, slide down a rope,
An Air Assault soldier ain't nothing but a joke.

CAPTAIN'S BARS

Twinkle, twinkle, little star,
Where did you get those captain's bars?

Please don't tell me, let me guess,
Two box tops and OCS.

FIRE MISSION

Fire mission, fire mission, coming down,
81mm on the ground.

FDC this is OP night,
I have the enemy in my sights.

FDC this is OP left,
Send me some of that silent death.

OP left this is FDC,
I have HDP and Willie Pete.

Boom,
Boom, boom.

AIRBORNE PATHFINDER

Up from the desert of Iraq,
An Airborne Pathfinder dawns his pack.

I walk all day, set up an LZ at night,
Airborne Pathfinder is out of sight.

Cobra one this is Pathfinder three,
I have a rag-head a pestering me.

Out of the sky through the darkness of night,
40 mike-mike rounds were raining down death.

It was prettier than a rainbow, horrified stealth,
40 mike-mike rounds were raining down death.

Cobra one, this is Pathfinder three,
A little bit closer, I'm almost free.

Well, I threw two grenades and I rolled to my left,
It didn't bother me that I was facing death.

I fired to my left, and I fired to my right,
Heads exploding, it was a hell of a sight.

Raider two, this is Pathfinder three,
I'm on the run to my new LZ.

Signal out, can you identify,
Green smoke rising in the sky.

Touch down, take off in six seconds flat,
Special operations, what you think of that.

Flying back home, swelled up with pride,
I'm gonna find a country girl and make her my bride.

MARINE CORPS

My corps

Your corps

Our corps

Marine corps

Is hard-core

SHOUT MARINE CORPS

C-130 rolling down the strip,
Airborne daddy gonna take a little trip.

Stand up, hook up, shuffle to the door,
Jump on out and shout Marine Corps.

MAMMA TOLD JOHNNY

Mamma told Johnny not to go down town,
Marine recruiter was hanging around.

Johnny went down town anyway,
To hear what he had to say.

Recruiter asked Johnny what he wanted to be,
Johnny said he wanted to join the Infantry.

Next Johnny woke in the pouring rain,
Staff sergeant said it was time for pain.

They jogged five miles then ran for three,
This is the life in the Infantry.

THE HAMILTON SHAKEDOWN

We are ready to fight, we're ready to drill,
We are ready to do the DI's will.

We are ready to lead, we're willing to follow,
We'll get our butts outta the hollow.

They say to err is human, to forgive is divine,
That ain't no part of that Corps of mine.

Gimme a hoorah
Gimme a hoorah
I said gimme a hoorah
Well alright right

WE ARE THE AIRBORNE

Hey, hey
Hey, hey

We are the Airborne
Mighty, mighty Rangers
We can fly higher than Superman

We are Airborne
Mighty, mighty Rangers
We are stronger than all your men

We are Airborne
Mighty, mighty Rangers
We can swim faster than Aquaman

We are Airborne
Might, mighty Rangers
We can climb faster than Spiderman

We are Airborne
Mighty, mighty Rangers

AIRBORNE RANGER

Send my regiment off the war, hey

C-130 rolling down the strip,
64 Rangers on a one-way strip.

Mission unspoken, destination unknown,
We don't know if we're ever coming home.

Stand up, hook up, shuffle to the door,
Jump right out and count to four.

If my main don't open wide,
I got a reserve by my side.

And if that one should fail me too,
Look out below Airborne Ranger coming through.

Hit the ground in the middle of the night,
Spring right up into a firefight.

GOING FOR THE GOLD

I don't know if you've been told,
But what I want is to get that gold.

One, two, three, four, hey,
Somebody, anybody, start a war, hey.

To pin that gold bar above my chest,
Will make me a part of America's best.

One, two, three, four, hey,
Somebody, anybody, start a war, hey.

OCS is mighty rough,
That's what makes candidates tough.

One, two, three, four, hey,
Somebody, anybody, start a war, hey.

When we graduate from OCS,
We'll represent Pennsylvania's best.

One, two, three, four, hey,
Somebody, anybody, start a war, hey.
Every day we run PT,

They run the sweat right out of me.

One, two, three, four, hey,
Somebody, anybody, start a war, hey.

Hard-core training makes us mean,
The meanest things you've ever seen.

One, two, three, four, hey,
Somebody, anybody, start a war, hey.

Combat ready every day,
That's how we earn out monthly pay.

One, two, three, four, hey,
Somebody, anybody, start a war, hey.

Fighting men go where they're sent,
Blood and guts help pay the rent.

PAPPY BOYINGTON

Pappy Boyington ruled the air,
The Japanese thought he fought unfair.

Twenty-eight zeros he shot down,
Pappy was the best fighter pilot around.

A Medal of Honor received with grace,
Pappy was the Marine Corps number 1 ace.

UP IN THE MORNING

Up in the morning half passed three,
First Maunakea bringing heat.

NCOs all around his desk,
Had a second lieutenant in the leaning rest.

First Sergeant, First Sergeant, can't you see,
You can't bring no smoke on me.

I can run to Maryland just like this,
All the way to Baltimore and never quit.

I can run to Washington just like this,
All the way to DC and never quit.

I can run to Alabama just like this,
All the way to Anniston and never quit.

I can run to Fort Bragg just like this,
All the way to Carolina and never quit.

KAMIKAZE KILLERS

Hey, all the way,
We run every day.

Airborne Rangers wear the tan beret,
Kamikaze killers and we earn out pay.

Jumping out of airplanes, running through the swamp.
Uncle Sam gets in trouble, Rangers gonna stomp.

Our minds are like computers, our fists are like steel,
If one don't get you then the other one will.

HIT THE BEACH

Up from a sub 50 feet below,
Up swims a man with a tab of gold.

Back stroke, side stroke, heading for shore,
He hits the beach and he's ready for war.

A grease fun, K-bar by his side,
These are the tools that he lives by.

How to kill a search team, a hostage snatch,
Airborne Ranger just a killing time.

He jumps through windows and kicks down walls,
Airborne Ranger just having a ball.

10,000 NASTY LEGS

Once there was ten thousand nasty legs,
Caught in the valley by the NVA.

The President was worried, he wondered what to do,
So he called on me and he called on you.

We jumped on in and did the job well,
Sent those commies straight to hell.

So if you have a problem, don't delay,
Call on the men with the tan berets.

Singing hey all the way,
Rangers lead the way.

One, two, three, four, hey,
Every night we pray for war hey.

One, two, three, four, hey,
Somebody, anybody, start a war hey.

Left, right, left, right, left, right, kill,
Left, right, left, right, you know il will.

SCUBA DOG

I had a dog his name was blue,
Blue wanted to be a SCUBA diver too.

So I brought him a mask and four little fins,
And took him on down until he got the bends.

Blue recovered and to my surprise,
He now dives to one thirty five.

RAVING MAD

Airborne Ranger raving mad,
He got a tab I wish I had.

Black and gold in a half moon shape,
Airborne Ranger, he's gone ape.

AIRBORNE (TO AMEN)

Here we go now

(Chorus)
Airborne

All together now

(Chorus)
Airborne

One more time now

(Chorus)
Airborne, Airborne, Airborne

All together now

(Chorus)
Airborne

Feeling good now

(Chorus)
Airborne

All the way now

(Chorus)
Airborne, Airborne, Airborne

SITTING ON A MOUTAIN TOP

Sitting on a mountain top beating a drum,
Beat so hard that the MP's come.

MP, MP, don't arrest me,
Arrest the grunt behind the tree.

He stole the whisky, he stole the wine,
All I ever do is double time.

Cause I'm hard-core,
Mean and lean.

Cut clean,
Looking good.

Ought to be,
Hollywood.

Every day,
One mile,
No sweat.

Two miles,
Better yet,

One, two, three, four, hey!
One, two, three, four, hey!

MY RECRUITER

My recruiter told me a lie,
You're gonna be Airborne, jump, and fly.

So I signed my name on the dotted line,
Now, all I ever do is double time.

Up the hill,
Down the hill.

Through the hill,
Airborne.

RANGER

One, two, three, four, hey
One, two, three, four, hey

Here we go,
Everyday

Gotta run
Airborne

Gotta be
Number one

Airborne
Ranger

R-running hard
A-all the way
N-never quit
G-gotta go
E-every day
R-Ranger

Hard core
Super trooper

Hard core
Paratrooper

JUMP, SWIM, KILL

Saw an old lady running down the street,
She had a chute on her back and boots on her feet.

I said old lady where you going to?
She said U.S. Army Airborne School.

I said hey old lady you're too darn old,
Leave that training to the young and the bold.

She said hey young punk who you talking to,
I'm an instructor at the Airborne school.

I saw an old man coming down the track,
He had fins on his feet and tanks on his back.

I say hey old man where you going to?
He said U.S. Army SCUBA school.

Whatcha gonna do when you get there?
Swim under water and never breathe air.

I saw a young man coming down the road,

He had a knife in his hand and a 90 pound load.

I said hey young man where you going to?
He said U.S. Army Ranger school.

Whatcha gonna do when you get there?
Jump, swim, kill, without a care.

MOMMA AND PAPA

Momma and papa were lying in bed
Papa rolled over to Momma and said

Give me some
PT

Good for you
Good for me

Get up in the morning in the morning with the rising sun
We run all day until the work is done

One mile
Having fun

Mile two
Good for you

Mile three
Good for me

Mile four
Want some more

Mile five
Keep it alive

Mile six
Great kicks

Mile seven
This is heaven

Mile eight
This is great

Mile nine
Mighty fine

Mile ten
Do it again

ARMY VS. MARINES

I don't know what I've been told,
The Marine Corps think it's mighty bold.

They don't know what the Army can do,
We are proud of our history too.

Out looks and style may not be smooth,
But you ought to see this Army move.

Look to your left and what do you see,
A bunch of jar heads just looking at me.

Shout it out and sing it loud,
I am a soldier and I'm mighty proud.

GROWING UP IN THE BAD LANDS

When I was young and growing up in the bad lands,
Momma said son what do you wanna be when you're a man?

Something with guts and a whole lot of anger,
Son you wanna be an Airborne Ranger?

When I was in school my teach said to me
Now that your done son whatcha wanna be?

Something with guts and a whole lot of danger,
Son you wanna be an airborne range

ROLLIN' ROLLIN' ROLLIN'

Rollin' rollin' rollin'
Ooh my head is swollen

Don't let your dog tags dangle in the dirt,
Pick up your dog tags tuck 'em in your shirt.

Rollin' rollin' rollin'
Ooh my neck is swollen

Don't let your dog tags dangle in the dirt,
Pick up your dog tags tuck 'em in your

Rollin' rollin' rollin'
Ooh my chest is swollen

Don't let your dog tags dangle in the dirt,
Pick up your dog tags tuck 'em in your

Rollin' rollin' rollin'
Ooh my hips is swollen

Don't let your dog tags dangle in the dirt,
Pick up your dog tags tuck 'em in your

UP JUMPED THE MONKEY

Up jumped the monkey from a coconut grove,
He's a mean mamma jamma you can tell by his clothes.

Rip stop cammies and a tan beret,
This little monkey was here to play.

Line a hundred sailors up against the wall,
He bet a hundred dollars he could whip 'em all.

Whipped ninety-eight 'till his fists turned blue,
Then he switched his fists and he whipped the other two.

UP IN THE MORNING

Up in the morning too soon,
I don't like it no way.

I eat my breakfast too soon,
Hungry as a hound dog before noon.

I went to the mess sergeant on my knees,
Said mess sergeant, mess sergeant feed me please.

The mess sergeant said with a big old grin,
If you want to be Infantry you gotta be thin.

Up in the morning too soon,
Eat my breakfast too soon.

I went to the mess sergeant on my knees,
Said mess sergeant, mess sergeant feed me please.

The mess sergeant said with a big old grin,
If you want to be Airnorne you gotta be thin.

Up in the morning too soon,

I don't like it no way.

I eat my breakfast too soon,
Hungry as a hound dog before noon.

I went to the mess sergeant on my knees,
Said mess sergeant, mess sergeant feed me please.

The mess sergeant said with a big old grin,
If you want to be Ranger you gotta be thin.

HEY LOTTY DOTTY

Left, left, left, your right left
Left, left, keep it in step

Oooh right a left
Left your right a left

Oooh right a left
I love to double time
We do it all the time

Hey lotty dotty
(chorus) hey, hey

Hey lotty, dotty hey
(chorus) hey, hey

Dress it right and cover down
(chorus) hey, hey

Forty inches all around
(chorus) hey, hey

Hey lotty dotty
(chorus) hey, hey

Nine in front and six to the rear
(chorus) hey, hey

That's the way we do it here
(chorus) hey, hey

Hey lotty dotty
(chorus) hey, hey

Standing tall and looking good
(chorus) hey, hey

We oughta be in Hollywood
(chorus) hey, hey

RUN, RUN, RUN

My grandpa is 95
He still runs and he's still alive

My grandma is 92
She likes to run and sing some too

I don't know what I've been told
If you never stop running
You'll never grow old.

YANKEES ON THE ROAD

92 yankees on the road
Running to the depot to get another load

One, two, three, four, hey
92 yankees on the way

Gotta get there by C.O.B.
I've got soldiers depending on me

One, two, thee, four, hey
92 yankees on the way

Been to the depot, gotta get back
My 1SG won't give me no slack

One, two, three, four, hey
92 yankees on the way

SUBMIT YOUR OWN CADENCES

Would you like to be in the next update to Modern Military Cadence®?

Submit your entry to
DunniganIndustries@me.com

If your cadence is used, you will be notified by return email and you will receive a FREE copy of the updated book.

We reserve the right to publish only those cadences we deem appropriate for our reading audience.